An Invitation to the Dance

The Awakening of
the Extended Human Family

Steven Jones

Welcome to the Dance!
With love & care,
Steven
x

An Invitation To The Dance

Steven Jones

First published 2010 by
Little Star Publishing
Essex, UK

© Copyright Steven Jones 2010

ISBN 978-0-9566895-0-4

Contents

Front Cover Artwork

As I was reaching the conclusion of this book I began to think seriously about the front cover. With the title being *'An Invitation To The Dance'*, how was I to possibly marry that image with the content of the manuscript.

After an amazingly short amount of time searching the Internet I was magically 'drawn' to something really quite inspirational. It was something that spoke to me on every level. A painting called *'First Dance'* by the renowned artist Natasha Sazonova.

After Annie and I had visited her website and seen the rest of her breathtaking portfolio and then read all about her thoughts and feelings about art, her life and the world, we knew, *without a single doubt*, that *'First Dance'* just had to be the cover for the book. We know that you will feel the same because Natasha's work is simply 'out of this world'.

From Natasha's website: http://www.artns.us/

Natasha Sazonova is a Ukrainian-American artist and illustrator. Her works have been exhibited in galleries throughout the Northeast, including the Museum of Art and Design in New York City. Natasha has also participated in a number of public art events with several world-famous Cow Parades among them.

Born July 22, 1978 in Poltava, (Ukraine), Natasha began painting as a toddler. At the age of four, she became one of the youngest students ever to attend the Fine Arts Studio at the University of Engineering and Architecture in Kyiv. Natasha started exhibiting her book illustrations at an early age and was planning on becoming a professional artist for as long as she can remember.

Natasha came to the United States at sixteen. She began her junior year of high school unable to speak English, and eighteen months later graduated in the top five percent of her class. She also holds a BA in Fine Arts from the University of Connecticut and Master's Certificate in graphic design from Sessions Online School of

Art & Design.

Natasha donates her talent and free time in order to give back to the country that became her second home and has participated in events where the proceeds of the sales of her works went to the Connecticut Children's Medical Center, Nutmeg Big Brothers and Big Sisters, Danbury Hospital Heart program and the Jimmy Fund.

As an artist Ms. Sazonova creates art works for private collections as well as commercial enterprises. She enjoys painting creative, non-traditional portraiture in a variety of styles ranging from whimsical illustration to collage art and surrealism. As an illustrator, Natasha specializes in editorial and book illustration. Since she believes that in order to be a good illustrator an artist needs to be a well-rounded person; Natasha spends a lot of time on self-education. She speaks four languages and keeps up with world literature, philosophy and politics. In her free time Ms. Sazonova enjoys cooking, gardening and playing with her dog, cat and four chinchillas.

Acknowledgement

A special note of acknowledgement goes to Jacqui Harvey who worked so creatively on the book. Her contribution went beyond the correction of grammar, punctuation and spelling and brought an additional level of insight to my thoughts and feelings.

I am extremely grateful to Winnie Durdant-Hollamby and Dr Michael O'Connell for "catching the ones that got away".

My special thanks go out to the people that have played such a vital role in my transformation:

Anne Cuvelier, Jack Weiner, Mary Weiner, James Weiner, Charles Foltz, Dr John Mack, Bob and Teri Brown, Budd Hopkins, Leslie Kean, Barbra Lamb, Ann Andrews, John and June Franklyn, Dr. Michael O'Connell, Mary McCormack, Debbie Kauble, Jason Andrews, Betty Balcombe, Bridget Grant, Judith Jaafar, Sapphire and Sean, Linda Cortile, Peter Robbins, Peggy Price, Donald Matrix, Mary Rodwell and Eva.

With love and light to my 'Little Star'.

To my fellow traveller in the wonderful shenanigans and spectacular monkeyshines of life, I praise your worthy choice of companion and for keeping me company on the long and winding road – I love you Reecey-Boy, my dear son.

To my beloved Annie – *words are not enough.*

Glossary

All subjects are noted in the main text with an asterisk *

Alien Abduction: the term 'alien abduction' is a generalized idiom used to describe the action of a supposed extraterrestrial entity in taking a human being apparently against their will and subjecting them, for often than not, to a series of physical and psychological procedures. New research suggests various alternative reasons behind this complex procedure that contradicts previous detrimental views on this matter.

Anthropomorphic: the attribution of human motivation, characteristics, or behavior to inanimate objects, animals, or natural phenomena.

DNA: a computer program is made up of sequences of ones and zeros. The way they are arranged tell the computer program what to do. The DNA code in each of our cells is very similar. It's made up of four chemicals that scientists abbreviate as A, T, G, and C. There are three billion of these letters in various combinations in every human cell. Just like you can program your phone to beep for specific reasons, DNA instructs the cell. DNA is a three-billion-lettered program telling the cell to act in a certain way. It is a full instruction manual for the human body.

Extraterrestrial:	in the common vernacular of the age, and specifically in relation to the subject matter of this book, refers to any living, sentient and intelligent being whose point of origin is believed to be beyond the planet Earth.
Experiencer:	a human being who has had direct contact with an Extraterrestrial. Also referred to as a Contactee or an Abductee, and, more recently, as a Participator.
Eschatology:	the branch of theology that is concerned with the end of the world or of humankind. A belief or a doctrine concerning the ultimate or final things, such as death, the destiny of humanity, the Second Coming, or the Last Judgment.
Greys:	who are believed to be a race of extraterrestrial beings whose skin colour is normally described as being a light shade of grey. Their height is between 3.5 to 4 feet, slight and fragile in stature with large black, almond shaped eyes.
Budd Hopkins:	the world's leading authority on 'Alien Abduction' and the founder of The Intruders Foundation in New York City. Mr. Hopkins has also published several highly successful and respected books on the subject.
Hypnogogic:	the state of being on the brink of sleep where the world of being semi-wake and falling to sleep overlap, a term first coined by the 19th century French scholar Alfred Maury.

Indigo Children: is a label given to children who possess special, unusual and/or supernatural traits or abilities.

John Keel: the late author and journalist (1930-2009) was one of the leading spokesman on the subject of UFOs and Extraterrestrials and was one of the first investigators to explore the direct relationship between these subjects and psychic phenomena. He is known to have coined the term "Men In Black" to describe the mysterious figures said to harass UFO witnesses after contact events.

Dr. John Mack: the late American psychiatrist, author and professor at Harvard Medical School was a Pulitzer Prize winning biographer and a leading authority on the spiritual and personal transformational effects of alien abduction experiences.

PSI effect: a term used in parapsychology to denote unexplainable or paranormal activity, such as telepathy, psychokinesis or Extra Sensory Perception (ESP) for example, and how and why they are produced.

John Spencer: a world-renowned authority on UFOs, alien abduction and paranormal activity and has written an extensive list of successful books on the various phenomena.

Pythagoras: Greek mathematical thinker, philosopher and theorist who lived from around 570 to 495 BC. Pythagoras was the founder of the Pythagorean brotherhood that proved instrumental in the formulation and development of western thought.

Screen Memory: a false memory planted in the mind of a person who has undergone what has become known as an 'ET contact'. It has been proposed that this action is carried out by the abductor so as to reduce the abductees' inability to accept the extreme nature of the experience. It is believed that Budd Hopkins created this term.

Jacques Vallée: an important author on the study of UFOs and their supposed extraterrestrial occupants and was later noted to promote the hypothesis for their inter-dimensional origin.

Introduction

"It is often wonderful how putting down on paper a clear statement of a case helps one to see, not perhaps the way out, but the way in."

Arthur Christopher Benson

Statement of Intent

Ever since I can remember, I have been an Experiencer*; that is to say I have been contacted by and have had an ongoing interaction and dialogue with an advanced race of beings that might best be described as 'out-of-this-world'. The common terminology for them is Extraterrestrials*, or ETs for short, although at all times throughout this book it remains just a term that I use as a way of identifying them, whilst not necessarily confirming their point of origin. They appear to be – in comparison to the human race – advanced both spiritually and intellectually; their knowledge and application of science is way beyond our current comprehension; their emotionally expressive range is very *different* to ours, although I would definitely not call it stunted or limited; and in most ways their appearance and style of conducting themselves is very *alien* to our own, which often causes confusion and lack of understanding and appreciation on our part. At every opportunity they are habitually misrepresented in modern human culture, although it has not always been that way.

This book is so much more than just the record of my contact experiences and how that has changed and affected me. In fact that element of my story is merely the outline of a much more complicated picture that is gradually being coloured in through my ongoing evaluation and learning of the world I live in. Hopefully through reading this book you will finally come to understand and appreciate, as I have, who *they* really are and what their purpose is in their communication with the human race.

And maybe, just maybe, we might even come to understand and appreciate who *we* really are.

This is a very positive story: I believe it will stir your heart and expand your mind; hopefully you will experience the joy, wonder and potency that I have felt through my contact with them. There have been moments of confusion and anguish on my part, but because of these episodes I have eventually found an inner wisdom and a sense of self-empowerment that I would like to share with you.

There was a part of me that didn't want to tell this story. It was the same part that had kept me frozen in life's headlights like a traumatised rabbit for the majority of my life; the part that had unknowingly coloured my views of the world and had been a major factor in influencing my day-to-day belief systems. This part was a misguided response to my contact experiences. It governed and it controlled, and not in a good way. Sadly, it took me over forty years to find out that it didn't have to be that way.

From five years old, I have had *contact experience*. It unnerved me and I was left feeling bewildered because it was never properly explained to me and there was so much going on in the 'real world' that either contradicted what was really happening or just plain denied it. To begin with I was happy for the contact experiences to be an honest part of my life, but after a while I found a way of hiding it away. Even though this part of me became compartmentalised, the impact remained the same; I grew from a happy-go-lucky little boy to a befuddled teenager to an anxious adult without consciously acknowledging where it was coming from. My experiences became an ingrained part of me, just like the nose on my face.

Finally, a sequence of events occurred in my mid-forties that helped change all that. After living with the proverbial nagging toothache for all of my life, I went to the dentist. Of course, it wasn't a pleasant experience, but if I had dealt with the problem a long time ago I believe I could have had a much happier life. In fact, I know I could have. So when the road forked, I took the road less travelled. The majority of Experiencers take the other fork; they have no other option: they are, after all, only human. The difference with me is that I finally noticed another option and by changing lanes and taking the exit ramp I altered my journey plan, if not my destination.

So, the part that was crippling me became the part that empowered. And that's the element that made me write my story. I believe that I have a responsibility to erect signposts for other seekers of the truth, so they too can change course.

The encounters of the Experiencer have always been the same; it's only our belief systems that differ and ultimately have an effect on the recall of the experience. Like the old adage about 'rose-coloured glasses', the lenses of an Experiencers' glasses can become twisted and distorted. By removing them altogether, they can, for the first time in their life, see the reality of what is happening. And although what is presented might stretch their belief system to snapping point, ultimately they will be the better for it.

Although some ET contact might differ in technique and purpose – after all it's a very big multi-verse with infinite possibilities – it is my belief that basically the objective is always similar, if not exactly the same. The trap we fall into most of the time is attaching our limiting anthropomorphic* view to the experience, whilst forgetting that we are dealing with – to all intents and purposes – an alien culture very different from ours. With all due respect to the human race, we really do need to accept that compared with other more developed species, we might seem just a little bit retarded in our response.

I didn't want to write a book solely about the details of the contact experience; clearly that has been done many times before, although Dr. John Mack* of Harvard University advised me that several aspects of my contact remained quite unique. What I have written is a chronological record of as many events that I can remember covering the last 48-plus years; events that I feel are relevant to the purpose of this book and which I have given the heading 'Experience'. In so doing I hope to identify a pattern that will play a part in our collective understanding of what has been happening in my life and why. I do believe that this is a joint project, so I am hoping that your participation will go beyond just the reading of my words. In fact, I believe that a vital component of the book's success will be the involvement of us all. There's a responsibility here for the reader that you don't usually find: buy this book and read it, but accept that you're signing up to something that goes beyond the norm.

In addition to these remembered episodes I also want to share with you the impact that they have had on me, and how they have shaped and influenced my thinking and beliefs. Without a definite comparison, I can only state that I firmly believe I would have been a very different person without them touching my world. Having said that, if I had published such a book five years ago, it would have been very different to the one I share with you today. For me, the series of proceedings that has led me here has been one of transition; growth through my experiences has been one of nurture, of learning, of discovery. It has been both a painful and a joyful process, but one that could not have been told until today. There are a few more missing pieces to this puzzle and I truly believe that it is you, the reader, who has those pieces in your possession. It will only be with your assistance that we can bring closure to this particular stage in our evolution and enable us all to, finally and successfully, move forward.

So, here's the pattern of my book: chronological descriptions of 'Experiences', which will include my reflective thoughts and feelings surrounding each specific episode. These are broken down into the following areas: *The Childhood Years, The Teenage Years, The Adult Years* and *Transformation*. Interspersed and sandwiched between the 'Experiences' are subjects and topics I hope will address and shine light upon the bigger picture. I have called these sections 'Food For Thought'. Additionally, I have recorded social events that have occurred over the later years, which have involved scenarios and people associated with the world of UFOs and ET investigation. These, *surprisingly enough,* I have titled 'Social Events'.

Expect more questions than answers in the earlier stages, but hopefully as we jointly move forward that might just begin to change. I'm not promising total enlightenment, but as this is *our* journey, together we should come away with a combined and increased awareness of where our future path lies because that is very much an important thread that is woven throughout the book.

What has happened to me has happened to countless others… believe it! This exercise has not been easy and I have worked extremely hard – emotionally, spiritually and intellectually – to complete this book and present its findings to you, things that ultimately challenged and rattled my belief system to its very foundations. Hopefully, by you re-treading the same route as I did, by-passing all the emotional trauma – *after all I've already done that for you* – you too will be left with the intellectual and spiritual challenges that these events present. The questions raised may push and shove you out of your comfort zone or you might take it all with a big pinch of salt. However, whatever your reason for reading this book, and however you choose to perceive it, I can assure you that the overall narrative will reach you on a much more profound level without you even realising.

Preface

What This Book Is *Not* About

This book is <u>not</u> about sensationalised depictions of flying saucers, little green-men from outer space or harrowing tales of alien abduction*. There are elements contained within the account that will relate to the description of advanced forms of aerial transport, the study and analysis of the very real and sentient occupants of these vehicles and their purpose in contacting and interacting with the human race, but it is most definitely <u>*not*</u> what the book is <u>*about*</u>. Whatever you find relating to the above topics is purely there as way of a map and to give the bigger picture both light and shade.

This book is, in fact, about you and me. It's about the man who sells us our groceries, the woman who sat next to you on the bus today, the policeman on the beat and the fire fighter saving our lives. This book is about our children's schoolteacher, the guy who sells newspapers on the corner of the street, the astronaut, the priest and the football player. It's about the politicians and the lawmakers, the criminals and the law-breakers. In fact, it's about all of us and it's my hope that it will be read by all of us, because I believe the facts and information contained within are applicable to every living human being and their future on this planet. In short it's about the ongoing evolution of the human race and the major leap forward we are about to take.

Where do I believe this book should be placed in a commercial setting? To the Internet traders, librarians and bookstore owners I urge you to consider the following: find a gap in the gardening section or the travel books; I suggest you make room in the crime thrillers, home improvement, cookery and history shelves. If you only take one copy, put it in with music, geography or science. What you must not do is hide it away with all the hard-to-find books in the dreaded section called *Paranormal, Occult, UFOs* or *Conspiracies*. Please do not put it in *Alternative Religions* either – god(s) forbid that you do!

Its home should not be in that misleading section called *Self-Help* with the books that continually tell you how to improve your world, whilst never allowing you to actually do so, because you're always so busy reading the following chapter or buying the next volume. And definitely, most positively, refrain from placing this title in the quagmire that encompasses all subjects beyond the safe and normal: the area ambiguously known as *New Age*.

If you need to pigeonhole this book in one specific area though, I humbly propose you place it in *Philosophy*. Not the philosophy you would hear from a Harvard or Oxford graduate however; what I am presenting here is the type of philosophical dialogue one would hear between work colleagues; exchanged by people whilst having coffee; over the dividing fence of two neighbours; shared by school friends in the playground. This is the *philosophy of life*, the common world if you will; the world that we actually live in, not the commonly inaccessible world contained within the minds and pages of academia. This is the philosophy of the universal man and woman; the people who speak a familiar language, not the exclusive language of the few. We all live in the same world together; why not speak in the same, understandable, language? Please do not think that I am in any way attempting to conduct some sort of 'intellectual bashing' with this statement; all I am working towards is everybody, and I do mean *everybody*, being able to understand, appreciate and discuss subjects that affect us all. In my experience I have discovered life to be very simple, so let's stop over-complicating matters.

I believe that it is relevant to this introduction for us all to be aware that Pythagoras* was the first man to call himself a philosopher; in fact, the world is indebted to him for the word philosopher. Before that time the wise men had called themselves Sages, which was interpreted to mean *those who know*. Pythagoras was more modest. He coined the word philosopher, which he defined as one who is *attempting to find out*. I consider that a much healthier position to be in. So we are all philosophers then, wouldn't you agree? For in my continued journey I haven't really come across anybody yet who *really knows*; but I have been blessed by coming into contact with an array of enlightened individuals who are *attempting to find out*.

What are the subjects we should be discussing? I suggest that we should start with Choice, Responsibility and Care. Are we all aware that we have choices in life? They may not always be easy ones, but we do have a choice. Where is our sense of responsibility? The one we should have for our actions and the words that we say. And where is the care for us – everybody that we come into contact with and the planet that we live on?

Within these pages you will read things that may or may not challenge your belief system: my contact and interaction with an intelligence that appears to be 'out-of-this-world'; my ongoing experiences with this intelligence and how it is helping me to evaluate my place in this journey we call life; my awareness of our potential as human beings and how we can all be a part of that joint success story for the future.

More often than not things are given labels in this world; I don't mean actual names, so we know what they are and can identify them, I mean superficial labels. The labels that are attached to people like me in certain circles are: abductee, experiencer, contactee... or sadly even *looney-tuney*! If a label were necessary, I would prefer one of *participator*. Here's the deal: I'm one of those people who claim to have been 'abducted by aliens'. In fact, to clarify that statement I would have to say that I have *an ongoing contact with a perceived non-human form of intelligence*. I also claim to have seen and been inside UFOs (metallic-looking flying saucers/modes of transport). I claim to have had experiences that could, if they really are true, quite easily dismantle our present understanding of the world that we live in.

Some of these events have even been witnessed or experienced by independent third parties, but that doesn't really matter, because I can't produce any so-called *evidence* that would satisfy a scientist – because they have no way to quantify this phenomenon, within their current paradigm anyway.

Am I looking for other people to believe me? Not really. What I'm really hoping for is this: whatever you believe about 'these sort of things', I would ask you to temporarily suspend it. Non-believer or even believer, please consider reading further with an open mind. What I'm writing about is hopefully much bigger and more important than 'little green men and flying saucers'. I truly believe that my experience, and the experiences of other people like me, actually means something important. In fact, it means something quite profound.

What I don't want this book to be is just another UFO book. There have been enough of them and on the whole they have been of a very high standard. Whatever their purpose was, it's been served. How many more times can we regurgitate stories of 'strange lights in the sky?' It's been 60 years since the birth of the modern UFO age and what have we really learnt? By looking at the UFO community (and being personally part of it) I reckon we haven't really learnt that much. Therefore, it's now time for a different set of questions. I genuinely respect a good deal of the serious Ufologists who have gone before, their investigations and the information they have documented. But are we really any further on than where we were in 1947? After all, our line of questioning has always been too 'human' to address such an 'alien' subject. And by 'alien' I am most definitely not referring to the accepted understanding of what the term 'extraterrestrial' (not of this planet) generally means, but just something that is so radically different to our current belief system that it is beyond our comprehension, and so by definition becomes 'alien'. Having said that, I do believe that the UFO experience is one of intelligent contact, but the language is so sophisticated and way beyond our – with respect to the human race – limited range of intellectual and spiritual hearing that we are deaf to its message. When it comes to books on the subject, with the exception of the innovative work of people like Dr John Mack*, Jacques Valle*, John Spencer* and John Keel*, it seems to me that some trees have died in vain.

I once heard a story about a tourist taking a vacation on the west coast of Ireland. I'm quite sure it wasn't supposed to be a true story, but just an affectionate caricature. The tourist was walking the country lanes in search of a particular village, but was having difficulty in locating it. He approached an elderly man who was sitting on a wall nearby and asked for directions. The elderly man removed his cloth cap, scratched his head and looked at the surrounding countryside apparently for inspiration. Eventually he smiled and said, "I know exactly the whereabouts of the village you're seeking."

"That's wonderful," the tourist said excitedly. "How do I get there from here?" "Now there's your problem," the elderly Irish man said frowning, "you can't actually get there from here."

Could this possibly be the situation we now find ourselves in with regard to the continuing search for understanding of the UFO enigma? In pursuit of that understanding, could we possibly have started from the wrong place? Will the questions that we continue to ask produce the answers we are ultimately seeking? Or is it now time to approach our subject from a different starting point? We are all human, therefore we measure intelligence against our own scale. Is this our only option though, or might we consider expanding upon our human way of thinking and try experimenting with alternative responses to what is being presented to us? In our contact with this extraordinary 'alien' intelligence, I believe we are being given – to coin a phrase – 'an invitation to the dance'. But, as yet, we don't know what that dance is or how to respond to the invitation.

I am now aware that any new line of questioning cannot be answered with scientific tools such as slide rules, microscopes or telescopes. As the free thinkers of our past once did, we need to consider opening our eyes just that little bit wider, so as to take in the bigger picture. Together we might begin to see the world from a slightly different angle and from that differing vantage point we might think about asking new and more informed questions. And that I guess, is what this book is *really* all about. It's not the details of my experiences that are important; they are just things that happened. But why they happened is important and what I could possibly draw from those experiences is vital in formulating the right questions. I will tell you about specific events that have occurred to me, but it will be for us all to think about what those experiences meant and what they are reflecting back to us all.

It is my belief that many of the leaders and controllers of the human race are just 'monkeys with computers', but that doesn't mean that we do not have the potential to become something else – something wonderful. It is my great privilege to have made contact with many human beings who have achieved this status already. The culture that we live in, the people that govern us, the media that dictates to us constantly dumbs down our innate potential for spiritual and intellectual development. We are constantly being told 'how things really are'. We appear to have forgotten how to question; we let too much slide. Where has our sense of responsibility gone, for our own actions and for the world that we live in? We have become complacent and lazy. Apathy has become our watchword (that's if we can be bothered – hah, hah!) Our supposed understanding, appreciation and dependency on science has outgrown our spirituality, or, as Martin Luther King so eloquently stated: "*Our scientific power has outrun our spiritual power. We have guided missiles and misguided men.*" That's very sad. It's also very dangerous. Imagine if you gave a machine gun to a monkey and set it loose in a local shopping mall. Or worse still, if you gave an atomic bomb to the wrong sort of human being in its present stage of development... Doh! Too late.

To make a comment on any area of the book or to contact me directly:

YOUTUBE   http://www.youtube.com/stevenjones57

E-MAIL   aninvitationtothedance@hotmail.com

BLOG   http://aninvitationtothedancersvp.blogspot.com

FACEBOOK Steven Jones

TWITTER   stevenjones57

Part One

The Childhood Years

"Come out, come out, wherever you are,
and meet the young lady who fell from a star."

Glenda, the Witch of the North, *The Wizard of Oz*

Experience: The Witch

London – 1962

O nce upon a time, in a magical land far, far away, there lived a young boy. Each day for him was a new adventure filled with the natural wonder and marvel of life. Some might say that his world was one of fantasy, but he knew differently. In fact, as he observed those around him he knew that it was *their* world, and not his, that was an illusion. He always knew that there were other children like him, but it would be many years before they were noticed. He often wondered though if any of the grown-ups had ever seen what he had seen or felt what he had felt. He thought that they probably had: before they had become grown-ups probably.

The geographical location of the magical land is not hard to find, but what went on there might be more difficult to pin down. The actual place was South London in England. The year was 1962 and I was the young boy. I was five years old.

I had a group of *very special friends*. It was in that year that they first came to play with me. To begin with they would only visit me at nighttime, when I went to bed, before I went to sleep. But as the years went by they would arrive at all different times of the day and night.

My most special friend, and the one that came to me first of all, I called *The Witch*. It wasn't because she looked like a witch, not at all. It wasn't because her behaviour was frightening either, for I was never scared of her. Actually, I was never scared of anything at that time; fear was an emotion that I would be taught to experience, rather than learn organically. The simple reason for calling her The Witch was because she could fly. I couldn't tell you how I knew she was female, because she certainly didn't look like any ladies I knew.

I always knew when The Witch was going to call because during the daytime I would get the 'tummy tickles' – as I liked to describe them – a feeling of 'butterflies flapping around my little tummy'. As I would potter around my little world, my stomach would start to feel strange and this would always herald a nighttime visitation. It was as if there was an invisible thread that was attached to my midriff that would produce a peculiar pulling sensation. When this occurred it was always accompanied by a tingling feeling that would eventually resonate throughout my whole body. On these occasions, unlike most other children at bedtime, my parents would not have any trouble persuading me to make that trip 'up the hill to bed-fordshire'. In fact, I would rapidly undress myself, don my stripy, soft and comfy, flannel pyjamas and jump enthusiastically straight into bed. After refusing the offer of any bedtime story, I would snuggle down under the covers as my Daddy kissed me goodnight and tucked me in; so tight at first I could hardly move.

Pulling the red candlewick bedspread up around my head and face, with only a small slit left for my eyes to peer out of, I must have looked like a knight-of-old staring out from his visor. I was in hiding. I was laying in wait. Filled with great anticipation, it wouldn't be long before my visitor arrived.

To give you an idea of what would happen next, I can liken it to an earlier memory from my childhood: my grandmother used to take me on the London underground system. After descending the wind swept, old-fashioned wooden-slated steps of the escalator, we would arrive upon the grubby and peeling, advertisement-plastered train platform. I loved to watch the train coming out of the tunnel and rushing into the station. Holding my hand very firmly my grandmother would allow me to stand on the edge of the platform and peer down into the darkness of the tunnel. Even before I could see the train coming there would be a change in the air pressure. With an urgent sense of potent energy hurtling towards me my ears

4

would 'pop', a tiny light would appear in the darkness and then the force of the train would enter the station. That's how it felt when The Witch would visit me.

In the darkness of my bedroom, lying still and silent, I would eventually become aware of a change in the air pressure. It would somehow push against me. There would be a peculiar aroma too, like the smell of the electrical transformer on my Scalextric racing cars when it overheated.

Straining to see in the blackness, I would eventually be able to make out a small, pinprick of light appearing across the room. It wasn't very bright at first, but would slowly grow in size and luminosity. It appeared to get bigger as if it were moving towards me from a greater distance than just a few feet away. A shiver of excitement would run up and down my spine. I remember that feeling so well. It's happening to me right now as I write these words.

Finally, as the light reached the size of a tennis ball, it would silently explode into a thousand tiny sparks. I would be momentarily blinded, and then, just as quickly, I could see perfectly well again. The room would no longer be completely dark. I was in the twilight with just enough light to see, just enough to realise that I was no longer on my own.

As I lay secreted beneath the covers, no more than a couple of feet away would be The Witch. Even today, after so many years, I can so easily conjure up the image of the very first time I saw her. I told myself it was one of my sister's toys, a lifeless, raggedy doll propped up against the wall. That was until she broke the spell by taking a step towards me. In that second she moved from my illusion into my reality. My acceptance of normality was shattered forever more. She became a living thing, but not any living thing that I had ever seen before. So although I was an innocent and naive young child, I certainly knew what was make-believe and what was not, and this was something else altogether. It is a fallacy promoted by adults that children 'do not know what they're talking about'. Even today, I conjure up the moment by imagining how I would feel now if an inanimate object suddenly came to life: a pale wooden mannequin with thin pursed-lips and staring-eyes or a rigid bronzed statue frozen and inert.

Like a butterfly emerging from its chrysalis, my head popped out of the crimson bedcover and I lay fixated in disbelief. The Witch's head made me think of the soft-boiled egg I would eat in the morning for breakfast: off-white, plump and smooth. Her eyes were like the surface of black balloons, stretched under the pressure of trapped air. I had never seen anybody who looked so skinny: her arms and legs were like the bendy pipe cleaners that I used to make Christmas decorations. I tentatively swung my legs out of bed and slid my tiny feet into furry *Fred Flintstone* slippers. When I looked back up, she was right in front of me. She was taller than me, but not by that much, she wasn't as tall as my Mummy and Daddy. I knew she was real, but in another way she was strange. I knew she was supposed to have a nose, but I couldn't see it. It was the same with the mouth: there wasn't anything there. That's why she was so thin, I thought, because she couldn't eat anything. I felt sad. She'll die soon if she can't eat, surely.

It was then that I *felt* her smile. And then I *felt* her talk. She said I should not worry about her not having a mouth, because she eats differently to me; *that* certainly made me feel better: first of all because she wasn't going to die soon, but mainly because she could talk and that meant she was like me, that she was normal after all and not a monster or anything like that.

Feeling more relaxed; I looked closer at what she was wearing. She didn't have a dress on like my Mummy. Her clothes were sort of shiny; they reminded me of my goldfish. They weren't gold in colour, but they were like the fishes' skin. Her hands were like her head – pale and soft – but there was something very different about them compared with mine. I had been able to count for years, and years, *and years*, and I knew that she should have had five fingers, but in fact she only had four. Her little pinky was missing, and her thumb was like the others – they were all long and skinny. I looked at my fingers and I could see they were different to hers: they had bendy bits, but she didn't have bendy bits on her fingers. But that was all right. I could feel her telling me that it was all right. I also noticed that her clothes went over her shoes, because her feet looked like goldfish skin too. I liked her clothes. I'd like to have clothes like her, I thought. I asked her if she wanted to be my *bestest* friend. I didn't understand why, but she said that we were already friends, that we had been friends forever. I couldn't remember ever seeing her before though, so I didn't know what she meant. But that didn't seem to matter either.

When I was a child I saw my life very clearly. I really didn't have to think too hard about anything, because my world was simple and straightforward; as the saying goes, 'It did exactly what it said on the tin', which is how I believe our world truly is today, behind the veil of illusion that is proffered without our knowledge. But as I got older and especially when I went to school for the first time, it was as if I was issued with a pair of spectacles with filtered lenses, so my world began to lose its full colour spectrum – its truth, if you will. And as time went on, the lenses became denser and the picture more distorted, which meant that it was even harder to recognize certain aspects of my world that had once been extremely familiar. Very scarily, I could no longer recognise who I had once been, until finally, it became virtually impossible to know what was real and what was not. The most unsettling aspect of all of this is that I was not aware of it on a conscious level at the time. It is only now that I realise my eyes have been shielded from the truth.

One of the welcome repercussions of my contact experiences since the year 2001 has been a change in how I see the world. And this has been a definite action on the part of the ETs, I believe, to fast track my evolutionary process. Through a course of realignment, my vision – optically and metaphysically – has stopped being processed through the artificial matrix that is our modern world of control and subterfuge. The veil has been lifted, which now enables me once more to see all that was clearly visible when I was a child. It is a 'return to arms', for I am now re-empowered with the full set of metaphysical toys that I was born with. And I am not the only person whom this is happening to; millions of adults around the world are experiencing this reawakening. Some are involved in ET contact and *some are not.*

In addition to this group of adults, there is a wave of children being born now and having the label of Indigo* attached to them.

These are children that display a much higher-developed range of senses: intellectual, mental, creative and metaphysical. Thankfully a high percentage of these children are born into home environments that nurture their abilities, whilst the remainder are hampered and restricted through being labelled as 'retarded' or 'difficult' or 'abnormal'. I was one of the first wave of Indigo children born in the middle of the last century and although not tagged as one in need of medication or therapy, I did spend the majority of my life with my abilities impeded by a world not ready to acknowledge the upgrading that was beginning to take place.

The percentage of children arriving now with heightened abilities is increasing dramatically; therefore it is of no coincidence that a frightened element of the medical profession is attempting to deal with something that they do not understand with inappropriate and unnecessary medication. These amazing children are being diagnosed as having a whole range of newly created *ailments* such as Attention Deficit Disorder, Asperger's Syndrome and Bipolar Disorder), when in actuality these children are completely normal or just super-duper normal you might say.

I'm sure you are far too intelligent to have not noticed, identified and acknowledged that certain aspects of the world medical authorities have plans – although not necessarily initiated by them – to carry out programmes of forced medical treatment on our children. I'm not suggesting that there is some evil force at work, although you might have come across a plethora of conspiracies to tell you differently and, in fact, it might actually be true, but it matters not a jot if it is. What matters is that the universe is lifting the veil. All of the older Indigos are being awoken and all of the new ones arriving are too sophisticated to be squashed down. And there is a reason for this: a purpose in the great design of things.

There has been a staggering jump in the percentage of children diagnosed with an apparent mental illness and treated with psychiatric medications. The Centers for Disease Control and Prevention in America said that in 2002, almost 20 per cent of office visits to pediatricians were for apparent psychosocial problems – eclipsing both asthma and heart disease. That same year the Food and Drug Administration (FDA) stated that in America some 10.8 million prescriptions were dispensed for children – they are beginning to outpace the elderly in the consumption of pharmaceuticals. And just recently, the FDA reported that between 1999 and 2003, 19 children died after taking prescription amphetamines – the medications used to treat ADHD. These are the same drugs for which the number of prescriptions written in America rose 500 per cent from 1991 to 2000. This is not acceptable.

Is something at work to stop our children from fulfilling their true destiny and for the human race to take the next step in their evolution? It's not just the medical profession either that's to blame: just look at the food stuffs that we poison our children with – the additives and chemicals, for instance. What about the pollution of the mind via poor television programming and violent computer games? Just take a look at the ineffective education system. This is not acceptable either.

There's something going on here. I'm not the only one who can see it. Children are not the same as they used to be. The thing is though, children are changing for the better and then we are giving them drugs to stop them from being different, when *different* is the key factor in all of this.

This change will happen, this evolution will occur, so we must not allow whatever forces there are at work to try to either stop or hinder this natural progression. This is just one part of the puzzle and without stepping back and looking at the bigger picture we cannot take on board how vitally important it is that we don't allow our children to be 'dumbed down'. Of course there are some genuinely ill people out there – children and adults alike – that are suffering from some sort of disorder that requires help, but please don't allow this minute percentage to disguise what is in fact something wonderful, something that needs our nurture and encouragement. We need to protect our children from this hammer blow of chemical medication, which is trying to 'fix' something we just don't fully understand. Let's see this for what it is: the human race is changing, it's been coming about for many years now, and we can no longer function in this crippling and retarded paradigm that we call our modern world. We are in the birthing process of a new human, a better human, a being that we have dreamed of becoming.

*"There are children playing in the streets
who could solve some of my top problems in physics,
because they have modes of sensory perception
that I lost long ago."*

J. Robert Oppenheimer, Professor of Physics, University of California

Food For Thought: A Box of Toys

It is my belief that when we are born we come into this world with a huge box of metaphysical toys. Within this box are many wonderful things for us to play with that are designed to enhance our lives. One such tool is the collection of our five basic senses: touch, smell, sight, hearing and taste. But there are other toys of a much more refined nature. One for instance is the ability to communicate our own thoughts and understand the thoughts of others without the need for normal speech. How often has a parent been caught out when wondering *silently* if their child wants juice or milk to drink, only to have their little-one chime back "Orange juice, please". What about when it is *wordlessly* considered by a parent 'time for bed', only to be informed by their reluctant child before the instruction is even issued, that they require *"Just five more minutes"*. This ability is generally referred to as telepathy*, which although unable to be proven by modern science (that's just their inability to do so by the way), is in fact experienced on a daily basis by every adult on a subconscious level. For very young children though, this ability comes naturally and is never questioned by them when it is exercised. As the child grows though, this apparently non-existent ability is squashed and finally turned off as the world around them refuses to acknowledge its existence. Why does this happen, you may well ask? Is there possibly

11

something at work to suppress this ability in the children on this planet? Was there a time when this ability was accepted as normal amongst us all? What other skills do young children have at their disposal that we do not find still intact, as they grow older?

I watch children and crave and envy their view of the world: seeing and experiencing so many aspects of it for the very first time and appreciating the beauty that is their life. They live without the jaded outlook that we all eventually acquire as we trudge through the matrix that we know as the material world. How do they see things? Can you remember? I have a sense of it, still trying to return to its previous intensity. Although I sometimes get a glimpse or a sense of the world through the eyes of a child – all too fleeting I'm afraid – and then it is very soon overwhelmed by the so-called 'real-world'. We are told to leave behind the world of our childhood and to embrace the world of adulthood. Who is up for returning to a vision and understanding of the world behind the veil? I firmly believe that young children are seeing beyond the extremely limited light spectrum that makes up our normal range of vision. Just watch a baby or a young child as they trace the movement of some unseen energy (intelligent or otherwise) as it moves supposedly invisibly across the room. You will hear people talk of the human 'aura' that encircles our bodies. There is no mystique attached to this – it's not magical; it is in fact just a visual manifestation of the electrical energy field that is given off by the human body – yet it is unseen within our limited range of vision. Stand in front of a baby though and watch them taking in the firework display that is our aura as it constantly changes in vibrancy and colour, reflecting what is going on for us at that moment: aspects of mood, health, feelings, emotions.

So if they can see this, what other wonders of nature are they experiencing? Do animals communicate? Of course they do, but do children have the ability to sense into that exchange of information? What is that exchange – is it electrical again? Or maybe it is transferred telepathically in some way (perhaps telepathy is electrically based). Who exactly is the intelligent species on this planet after all? I know for a fact that my dog used to place thoughts into my head when it wanted food; I'm certain she knew what she was doing when she did so, but can we consciously carry out that action with ease or awareness?

Can we see X-rays, television signals or radio waves? No, but they still exist don't they? How do we know that? Because the majority of us trust the scientific community to tell us what is real and what is not. If I began to believe that the invisible signal sent out via my satellite to my television receiver did not exist, would it mean I wouldn't be able to watch my favourite programme? Of course it wouldn't. I don't have to believe in something or understand how it works for it to be real and have an affect on my life. Just because you don't believe it doesn't mean it isn't true! The very sad thing is though, that if a child is told enough times that the world can only be experienced through the five accepted senses then that will very soon become the only world they experience. All the other 'toys' will eventually become unavailable for them to play with. Are our children being dumbed down? Is this being carried out by the educational system, by television, by the media? Of course it is. In your heart you know it's true. In fact, we are all being dumbed down, but accepting it as a reality is far too scary, isn't it?

"There is a garden in every childhood,
an enchanted place where colours are brighter,
the air softer, and the morning more fragrant than ever again."

Elizabeth Lawrence

Experience: Picnics on Clapham Common

London – 1962

In the same year that my visitations began, I had yet to begin any sort of school and because both of my parents worked full time, I was cared for everyday by either my grandmother or my great aunt. Although I loved both of them the same, my most treasured memories come from the time I spent with my great aunt – Auntie Katie. Being with her would normally mean a day filled with wonder of one sort or another.

My memory of her is crystal clear: she was as different to other grown-ups as I was to other children. Nothing was ever said to confirm this feeling, but I could see it in her eyes every time she looked at me. She never spoke to me or treated me like a child, so I never behaved like one. We had an understanding.

My favourite adventure was when we made up a packed lunch and ventured out to a place called Clapham Common in southwest London. It was a region of open grassland and woods. Although later on in life I would realise that the Common was relatively close to where my aunt lived, at the time it always felt like an extensive trek to get there. We always had a preferred location in mind when we left home, which was a small grassed area to lay out the picnic things and a dense coppice of trees nearby for my play area and entertainment. I was always impatient to consume the selection of snacks that Auntie

Katie laid out neatly on a blanket for me, because I was keen to go exploring. The liquid refreshment to accompany the food was always warm milky coffee. Whenever I smell the aroma of a frothy latte today it takes me back to those magical days.

Once the food was eaten, an extremely familiar routine would take place: any remaining food stuff would be placed back in the basket, my aunt would then roll her coat up as a pillow and she would lay down with her head resting upon it. Remarkably, she would then proceed to have a nap. Her final instruction before drifting off to sleep would be, "Be a good boy on your adventure and come back safely". When I grew to be a man and had the opportunity to view the situation through the eyes of an adult, a parent even, it was only then that I saw the absurdity of the situation – a loving and responsible grown-up carer letting a pre-school child wander off on his own, whilst they lay down and slept. Be it then, or especially in this day and age, it just wouldn't happen, not unless there had been some kind of hidden agenda or third party influence at work. At the time however, neither I nor even my aunt would have thought of questioning this bizarre and ridiculous situation.

I walked towards the coppice of trees with a sense of purpose; I was going to *visit my friends*. After a few minutes of carefully navigating my way through the wooded area I came to the clearing that was my destination. I had located this point on many occasions and truly believed that I could have found it with my eyes closed. It felt as if there was an invisible cord that somehow attached me to it. I always felt a tingling rush through my body as I stepped forward. It reminded me of when I had rubbed a balloon against my head at a party and all of my hair had stood on end. There in the clearing stood what I referred to as 'the hospital'. In height and width it was about the size of a London double-decker bus, the sort that I loved to ride on, especially when I was allowed to take the front seat on the upper deck. Of course, it didn't look like a bus or it would have been a bus, it was just big in that way. It certainly didn't look like other buildings I'd seen before: this one had rounded edges and was extremely smooth – it wasn't rough and made of bricks that's for sure. Its surface was dull and grey and looked like it might be made of metal. It looked peculiar, but that really wasn't very important to me, in fact I never gave it much thought at all. What went on inside was always of more interest to me.

The hospital had a row of little round windows running round the middle of it outside, about half way up the wall. You couldn't see into them, they were too high up. I wasn't sure if they went all the way around because I never looked. Right in the middle of the front of the hospital, a little bit above floor level was an opening that had a ramp leading up to it and a thin white curtain across it that used to flap in the breeze. It would wrap around my face whenever I walked through it and I always had to untangle myself. It felt sticky against my skin. I walked up the ramp, through the curtain and into the opening. Recalling the memory when I was older, I was fascinated to acknowledge that the space I entered into was always comparatively more expansive than it should have been, considering the relative size of the structure from the outside. But at the time this irregularity did not register. Several nuances of this nature were apparent to me at a later date, but as a young child they didn't even raise a question and when you think about it, why should they really? I had no concept of why it would be impossible for the interior of the building to be greater that the apparent confines of the outside walls. Children accept the world based on their limited understanding of it.

The space inside was very bright and white with a high domed ceiling. Around the circular wall were the open doorways to four other rooms or corridors – they were always too dark inside to enable me to see their interiors. Also, I couldn't see the little round windows that I'd noticed from outside and couldn't work out why. But that didn't matter.

After my very first visit I always knew what to expect. After a short while after entering, several rows of small children would file out of each doorway – about six at a time – walk towards me, and then sit cross-legged on the floor. One of the main reasons I always considered myself to be in a hospital was that these children didn't look very well. They were all about the same size as me, but were very fragile in appearance. There was a lacklustre to their movement that made me feel sad; they seemed confused and unsure of every move. They were all dressed in the same way, plain white smocks that trailed on the floor, just like oversized nightgowns. But it was their faces that troubled me most of all. Their heads looked like ping-pong balls with wispy strands of hair stuck on here and there. The eyes were dark and bigger than mine and their skin was very pale. Once again, when I was older, I would see something that would remind me of these feeble looking little characters: cancer victims who had undergone chemotherapy. I understood that although I was profoundly different from my sickly looking friends, I liked them all the same. Although they seemed lethargic and not at all communicative, I enjoyed playing with them nonetheless, because I knew somehow that I was helping them; that I was doing something special and clever – *something that only I could do!*

Teaching them the rudiments of playing games remained a struggle from visit to visit. Either they just didn't understand me or had a problem retaining information. It made me frustrated and grumpy sometimes, but I persevered. After all, *I was in charge.* When I showed them how to play *Ring-A-Ring-o'-Roses*, it took me forever just to get them all to stand up at the same time; then to make them all join hands in a circle; and finally to move around as I sang. They never sang though, however much I encouraged them to do so. Finally, when it came to the part where they should 'all fall down', the instruction was ultimately carried out, but their actions were mechanical and they dropped down as though they were heavy bags of potatoes. The one game that I was never able to play with them was 'It'. This was the game where one person chased round in pursuit of everybody else and when they came within touching distance they would reach out and place their hand on them, whilst squealing 'You're It!' It would then be the job of the 'touched' person to begin the chase again. It didn't matter how many times I showed them how to do it, they never understood what I meant; they just stood there staring at me, appearing to be waiting for further clarification. Eventually I just gave up and moved on to something else. It was painfully apparent that the kids never laughed or even smiled. I thought it must be because they were all so sick. It never stopped me from trying to play with them though; being judgmental with regards to the behaviour of others would be taught to me in the future by grown-ups. Because they all looked the same, as far as I was concerned that is, it took me quite a while to realise why they probably didn't remember how to play the games on each visit: maybe they were different sets of kids each time! Now and again, I felt that I recognised one of them from the time before, but it was more a *sense of knowing* than actual physical recognition.

Eventually, after what seemed like many hours of play, my very special and unique little friends would just break away from me and begin to file back into the open doorways in neat lines. I never had an inclination to follow them, as I somehow understood that I wasn't supposed to. As the last of them disappeared into the darkness of the doorway, I turned and walked away, knowing in my heart that I would return as I had done on numerous other occasions.

Getting back to the picnic area, I found my great aunt waking from her nap and unfolding the coat that had been her pillow. "How was your adventure today?" she would ask. "It was really great", I would enthuse, as we began our journey home. Although I would share all the details of my exciting activity with my great aunt, I would never reveal those secrets to anybody else; somehow I knew that was all part of the game.

I truly wonder at the ability of young children, as I was then, to accept and embrace the differences of others. It fills my heart with such joy to allow the memory to resurface and rekindle. As the weight of the world is allowed to sit squarely upon our shoulders, we walk with a stoop that reflects in everything we do. But it doesn't have to be like that; we allow it to be like that, but we can make a change. We have to try to become aware of our actions and the words that we share with others, and in that awareness we will become empowered; that will ultimately become our new habit, our new way of living. It won't be easy – nothing challenging and ultimately rewarding ever is – but my ability to practise what I preach is increasing daily. It's all about being 'awake' and not meandering through our waking day metaphorically crashing into people with inappropriate words and unacceptable actions. If we had the luxury of a 24-hour video recording of our daily life, we could easily pinpoint what went wrong – it would quite literally stand out a mile. But that is not the real world, so what we have to do is 'stay awake' – try it for ten minutes at first (it will feel like an hour to begin with), but stick with it. Then gradually extend the process: *stay awake* to what you are doing and saying. Change one habit for another! After all, our unacceptable behaviour now – the one where we hurt and upset others, either without thinking or realising what we are doing (ultimately impacting on ourselves) – was a learnt behaviour; we were all nice guys to begin with, were we not? Try and make contact with that pure child that you once were; the one that did not judge; that did not discriminate; that did not lash out and hurt others with inappropriate words and actions. I can remember that very pure feeling of acquiescence; that feeling which is ultimately the essence of pure love.

Are we taught to differentiate amongst ourselves and then to act upon that awareness with repulsion and rejection? At what stage are we taught these prejudices? Does it come through at any one point of our education or is it an amalgamation of influences: parents, family, siblings, friends, school, television, comics, films, and books? Why does it appear to be so naturally acceptable in human beings to reject anything that is different? Could it be innate – just a natural element of our make-up – or might it have been 'placed within' our DNA* at some point as a form of protection against that which is different to our species? Further exploration of that concept will follow later in the book.

I will say this one thing – and it could be seen as a weakness, but by sharing it with you it becomes a strength – even now, after nearly 50 years of my contact with a race of highly evolved beings who display no concept of rejection through appearance, and despite their ongoing education and reawakening of my psyche to take responsibility for my words and actions, I *still* struggle to work via my heart's energy in all things. I guess it's difficult to let go of the negative programming after so long; it must go so deep into our psyche after having it pounded in on a daily basis. But we must never forget that we are 'beings of love', so we just need to lift the veil that conceals our true nature and see who we really are; anything else is just a veneer, after all. I believe that the universe is encouraging us to look at ourselves and if we can accept what we see – even if we are initially disappointed by some behaviour – there is a bright light that shines out from our hearts, a light that is so true and pure that it will put anything unacceptable into the shade and without access to that light all negativity will surely die.

I am becoming that five year old child again – it won't happen overnight – but as each day goes by, that feeling of love, acceptance and awareness is seeping back into every molecule of my body and I am gradually becoming the old me, not a new me, but a very dear, erstwhile, friend.

"And those who were seen dancing were thought to be insane by those who could not hear the music."

Friederich Wilhelm Nietzsche

Food For Thought: For All Experiencers

Before I officially began work on this book and was still attempting to identify what its real purpose should be, several people offered me counsel to help reach a constructive conclusion. It was quite simple really, they said: *"It's for all the other Experiencers who haven't, for whatever reason, found a voice to share their feelings about what has happened to them. It's for the people who still live in fear because of their Contact and for them to understand, through your example, that one can become free of those disabling shackles. It's for the singular Experiencer to let them know that they are not alone; that there are millions just like them, all an integral and vital part of this weird and wonderful coalition. It's to help them appreciate that although there does not appear to be any defining answers, purely in our legion we are empowered to formulate the right questions. "*

To all those counsellors, I say thank you for the validation.

On the flip side of the coin, it is also for the people who haven't had any form of contact experience. Hopefully, through reading about my experiences and my broader thoughts on the subject, these *inexperiencers* might be able to view our circumstances from a more informed vantage point and form a more supportive opinion. After all, this narrative has never been about convincing anybody, but rather to help educate and inform (if that doesn't sound too patronizing).

So draw ever closer my fellow Experiencer, but please allow room for everybody else who would like to listen.

I have known profound fear because of my contact, but I have come to believe that by inadvertently filtering the experience through an imposed and debilitating belief system, the memory of what occurred became something other than what it originally was. During my childhood the recollection of the contact was not tinged with the belief that I would come to any harm. It was only as I moved into my teenage and early adult years that each experience was *remembered* with an element of distress and anxiety.

It's a funny old thing isn't it: experience and memory. For although they are forever connected – sharing the same space as it were – they can be complete strangers. My experiences have been like that – not all the time and certainly not now – but they have been at times strangely disjointed. So consider this: although our contact might be remembered as a traumatic event, it may not have actually been that way when it was first played out and it certainly doesn't have to be that way in the future, and if you are unable to believe that now, at least know it in your heart as a possibility.

How might one attain that new belief and make it a working reality? I began by talking to the ETs in my head and by creating positive visualsations in my mind. As I lay in bed at night, drifting in and out of clear consciousness – in a state that has acquired the term '*hypnogogic** (thank you RJ) – not fully awake, but also not in the true dream realm, I would begin to compose intentions. I wanted to convey to the ETs that I required my contact to be of a positive nature and for it to be remembered as such; that I needed to be consulted when it came to their ongoing agenda and I wanted to stop feeling as if I was being used and abused (or at least remembering it that way). I began to visualise positive communication with the ETs. Although I might vary the wording and the backdrop, the intention was always the same: I needed encouraging and conscious contact; participational and influential interaction; and respectful and considerate behaviour. I repeated this every night for months and in so doing I began to believe it possible. Through this repetition I began to remember past events differently, as if the fear element began to evaporate. I also started not to dread their coming; if anything, I began to invite their contact. Over a relatively short amount of time everything changed, but the most important element was that I no longer felt scared and this rolled over into my day-to-day world.

Throughout this period of transformation I had no obvious or apparent contact with the ETs. I didn't know for sure if my requests were getting through, but what I do know is this: that the next time I was *taken*, because of my new way of considering my part in events, I was able to not only challenge them, but to effectively alter how my ongoing contact would be in the future. They *appeared* surprised at how I responded to them and I was directly asked what had happened to me since we had last met. So, by definition, that would suggest that they had not heard my affirmations (although I find that very hard to believe) and what was to follow – my empowerment in events – was purely down to the reframing of my own belief system. Having said that, I also believe that something was then done to me by the ETs – organically or metaphysically – that enabled me to have access to the memories of all previous contact events and for them to be remembered for what they had truly been at the time – or had not been, would be a better way of saying it – for the memory returned without any degree of fear. Clearly that must have played a crucial part in how I now saw them.

Since that time – although it has not been said outright – I have a sense that they had been waiting for me to make this breakthrough; but it was always going to have to come from me; they could not induce or suggest this change of mind; I had to make this advance under my own volition. I believe, in my heart of hearts, that they are waiting for all of us to take that leap of faith; to remove the filters that colour our contact experiences with varying tones of fear, and in doing so our outlook on life will change. I no longer read tabloid headlines or listen to depressing news broadcasts with dread and anxiety, because I know that, firstly, it may not even be true, and secondly, I believe that we can do something about it; we don't have to live in fear anymore, just because we are told to. The curtain has been pulled back to reveal that the scary 'Wizard of Oz' is in fact a scrawny little playground bully, whose real control and manipulation of us is only in our minds.

I encourage all Experiencers – of ET contact and of life in general – to make that leap of faith. Stop being frightened. If you can make that life change, the veil will lift to reveal the truth – the one truth

– that without fear we become our true selves and can no longer be manoeuvered into what to think and believe. We will begin to make up our own minds.

I am that example. It can be done.

Do you want to be afraid anymore? No, I didn't think so.

We will discuss this matter again, in greater detail, further on in the book and I will take you through a 'working and practicable example' of how to achieve this empowering transition.

"The great awareness comes slowly, piece by piece.
The path of spiritual growth is a path of lifelong learning."

M. Scott Peck

Experience: What's Going to Happen?
Something Wonderful!

London – 1963

The year 1963 was a year of extreme polarities: the United States of America saw their governor for Alabama vowing "segregation now, segregation forever" and attempting to block the entrance to the state university to African-American students. Yet, at the same time, the U.S. president, John F. Kennedy, was categorically stating that segregation was "morally wrong".

Ironically, Dr. Martin Luther King was arrested and jailed in May of that year after campaigning for equal human rights, but then successfully marched on Washington in August and delivered his 'I Have A Dream' speech before a crowd of 200,000 gathered at the Lincoln Memorial. It was a time of hope and heartbreak, ripe with the promise of beginnings and scarred by tragic endings.

On 22nd November, John F. Kennedy was assassinated in Dallas, Texas. I can remember the television programme I was watching being taken off the air – only to be replaced by a grave looking lady standing next to a man playing a piano, whilst she sang mournfully. In my opinion, we still have insufficient appreciation or real depth of understanding for the lost opportunity we suffered that year.

I was six and three quarters.

As with a lot of boys my age, I had an unexplainable fascination with things of a mildly gruesome nature. Therefore, the fact that

I experienced semi-regular nosebleeds did not disturb me in the slightest. Although I did not encourage them to occur, when they did I was always fascinated by the process. Regrettably, they would often take place during the night whilst I was asleep and I would find the traces on my pyjamas or on my pillow when I awoke. Sometimes though, something would wake me and I would be able to mop the stream up with my handkerchief for closer inspection. On one occasion, it happened during the daytime and I was able to turn the clean, crisp, white implement for blowing my nose into an exquisitely sodden crimson banner and then share this exhilarating experience with my mother.

I had been taken to my Great Auntie Katie's house. Whilst she and her grown-up son – my second cousin, the man we called Uncle Kenny – and my mother were engrossed in conversation, I went exploring the house. Although I was already familiar with every room, it was always an adventure to investigate as if everything was being newly discovered. I made it into a game. The staircases were mountain tracks; every hallway was a secret corridor; and all the cupboards were magical entrances to caverns laden with treasure. As I reached the very top of the house, just inside the bathroom, was the entrance to the attic. I heard a shuffling noise behind the bathroom door. Pushing it open and stretching up to turn on the light switch, I saw the shadow of somebody on the wall to the left of me. Peeping slowly behind the door I was greeted by a very familiar face: it was The Witch. I didn't feel frightened or surprised to find her standing there. In fact, I had had one of those feelings that she had been close by for most of the day. The very familiar tingling in my stomach had been a sure sign of an impending visit. I smiled. And then I *felt* her smile back. She wanted to play a game, I heard her say; it felt like a feather stroked across my forehead when she did so. "Okey-dokey", I replied eagerly. "You have to close your eyes," she instructed me. "What's going to happen?" I asked, not caring, but just inquisitive. "*Something wonderful,*" I heard her words whispered in my head.

I opened my eyes after what seemed like a couple of seconds, impatient with waiting any longer, to find out what was going to happen next. The Witch wasn't there. Strangely enough, I was no longer at the top of the house anymore; I was now out in the back garden. Looking around me, I couldn't see The Witch anywhere. I was quite alone. The Witch's tricks were always so special I thought, but I wished that she didn't keep disappearing like that!

Although I had only just had my lunch, I was now starving. It felt like forever since I had had anything to eat. I sniffed. *I sniffed again.* Wiping the bottom of my nose with my hand, I discovered that it had started to bleed. *Terrific*, I thought. Excitedly I took out my neatly folded clean white handkerchief and began to eagerly blot the slow trickle that now reached my upper lip. I knew that I should never force it; my Uncle Kenny had told me I would 'bleed to death' if I did. Tilting my head back slightly I squashed the now crumpled hanky against my nose. After a few moments I released the pressure only to find that the trickle had turned into a constant flow. There were very few white areas left on the handkerchief, so I quickly refolded it to accommodate the ongoing stream. For the first time ever, I began to feel concerned that it wouldn't stop. It didn't feel interesting anymore. It felt scary. Flinging open the back door, I rushed in from the garden calling out for my mother. I found her sitting in the kitchen still engrossed in her conversation. As she looked up, I produced the handkerchief, now completely saturated with blood. "Oh for god's sake, what have you been up to now? You've been hiding for hours, we couldn't find you anywhere!" my mother complained in a clearly unsympathetic tone. As I began to explain, something small dropped from one of my nostrils and I caught it in my hand. Peering at this mysterious projectile, I saw a small round ball, orange in colour and slightly bigger than the head of a pin. Rolling it in between my fingers it felt quite hard. Offering it up to my mother for closer inspection, she grimaced in obvious revulsion. "Don't be so disgusting, throw it away immediately."

In the bathroom I continued to prod the small orange ball with various items I used as tools: a toothbrush, a nail file and even a razor blade that I had discovered in the cupboard underneath the sink. Nothing seemed to have any impact on it. It wouldn't cut in two, however hard I tried. It wouldn't even bend or squash in any way. I did consider concealing it in my pocket and having another go at it when I returned home, but I knew that my mother would be mad with me if she found out. So, for the first time in ages, I did what my mother had told me to do. As it spun uncontrollably round and round in the swirling water of the flushing toilet, I did wonder if, just maybe, it had been a present from the Witch and that possibly I was throwing away something wonderful.

Although I was never to find another little orange ball, the nose bleeds continued on and off for years. Without ever asking The Witch directly, I was somehow aware that she already knew all about them.

Replaying this childhood memory through the eyes of an adult, it is clear to me, based on things that were still to come, that the event of that day was a contact experience and that I was probably removed from the house for a period of time apparently longer than I could have imagined. I would also suggest that, because of the object that fell from my nose during the bleeding process, I must have gone through a procedure to either implant or remove the said item from my person. The fact that it came into my possession at all would also suggest that the procedure must have gone wrong in some way.

It has been suggested that the 'implants' the ETs secrete about our body are devices to track our location. I have never believed this to be true. In fact, I believe that each human body gives off an individual energy signature that enables those who have the ability to trace our whereabouts whenever they want to. I believe that the 'implants' are there to register and gently influence certain aspects of our bodily functions – physically, mentally and emotionally. These actions are for our benefit; they are most definitely benign. However, the technology behind these devices might now have fallen into the hands of those who do not necessarily have our best interest at heart. Therefore, their ability to manipulate our behaviour on many varied levels – sometimes with or without the need for implanted apparatus – is rather more malignant and self-serving; in short, the need to try to control our thoughts and actions.

If I have said it elsewhere in this sequence of events, I apologise for the repetition, although it is worth repeating: a very wise counsel of mine once said, "Don't be worried about little green men from Mars invading the Earth and taking over. The menace you need to be concerned about is much closer to home. There are those that are much more human that would have you suppressed and malleable to their intent. *Sometimes the scariest monsters of all can be your fellow human beings.*"

I believe that the ET contact experience is something wonderful – if we chose it to be. Based on our individual belief systems, we have the ability to turn one event into many experiences. On the whole, *contact,* is all the same, but we all process it differently. If only we were able to see and feel the same. I appreciate we are individuals, but after all, there is only one truth. If only we were all capable of experiencing the truth as one. I believe that we are working towards achieving just this and that our ET contact is a vital part of that process.

"Love is what we are born with.
Fear is what we learn."

Marianne Williamson

Food For Thought: Learning to be Fearful

What so-called 'skills' do we develop that we do not have at birth? How about prejudice? Does a very young child judge you by the colour of your skin or by what part of the world you originate from? Would it be possible for an infant to pass judgment over the belief systems of various secular faiths? The answer to these questions is obviously no. Sadly, our children are sometimes *taught* to be insular and intolerant, be it through their peers in the playground, from the inherited ignorance of their family or by the television programmes they watch. In truth, a lot of children receive negative programming from the world they live in.

Where does a child's fear come from? I'm not talking about the fear of walking out into a main road without looking – that's just good education. What I'm referring to is the fear of the world around them; the fear of something different to what they know. Our children are taught to fear the opposing beliefs, views and opinions of others. Through television programmes and newspapers, our children are told to fear a different skin colour or an alternative belief system. If it's different to what you believe in or what we have told you is acceptable, then view it with fear – that's what our children are hearing. No wonder the opposing factions around the world – political, devout or otherwise – continue to put all their efforts into either disproving the beliefs of others or attempting to annihilate them from the surface of the planet. Where does this come from?

We appear to be tribal in our nature or rather we appear happier when split into basic tribal groups, but does that come naturally or is it taught behaviour? From the street we live in, to what country we reside in; from which local football team we support to our national team allegiances; from social group affiliations to international submissive belief systems; from basic life convictions to colour and creed; from the earliest opportunity we are separated, manipulated and controlled. But sadly, through this enforced segregation comes fear. We are systematically detached from our natural state of Love, of being together and not being scared of one another. Love is the natural human condition. Believe it or not, I know it to be true. And if you are totally honest with yourself, you also know it to be true.

Why don't we all look the same? Yes, we appear to all come from the same mould, although we differ in many ways: skin and hair colour, the shape of our features, height and a million other little aspects of our appearance. We have our differences, but we have the ability to overcome them and not allow them to destroy us. To achieve this natural state we only have to acknowledge and exercise our innate intelligence and spirituality; in other words, we just have to be ourselves.

So what about the animal kingdom? As far as I can see segregation comes about as a response to the threat of being eaten, which I suppose is the nearest some animals come to what we might classify as violence. Although I have witnessed some species aggressively protecting an area of ground, that is usually related to food or a watering hole, for breeding purposes or protection of young, I have never seen acts of violence being carried out for violence's sake (apart from one of the opening scenes in the fictional movie *2001: A Space Odyssey* – but they were trainee humans after all!) There might be those who will 'nit-pick' my comments by showing abstract examples that could challenge my theory, but let's look at it in black and white: humans kill each other to suppress, to steal, to create fear, to control; animals kill through the natural chain of survival. Animal actions are natural in these circumstances, whilst human fear-based behaviour is learned. I do wonder how cavemen really lived? How was their social structure organised? Did they go about killing each other? Were they tribal in a segregated sense? Is my way of thinking innocent and naive? Or is it right on the money?

There is a delightful family anecdote of ours that perfectly dovetails with my overall thoughts on this matter. It refers to the time that my stepson, Glynn, attended his opening day of infant school. Annie, my wife, told me he was about five years old at the time; bright as a button, unknowingly mischievous and with quite a unique and untarnished view of the world around him. As Glynn sat there at home ruminating after his first day, Annie busied herself preparing his teatime meal. "I made a new friend today" her son said kicking off his already scuffed new black school shoes. "That's nice darling, what was his name?"

Glynn thought for a moment, then said, "I think it was something like Toki Singh."

"Oh, I see. Is he an Asian boy sweetie?"

"What do you mean?" came Glynn's reply, screwing his face up in puzzlement.

"Well... does he have brown skin for instance?"

Glynn thought longer this time before he answered.

"I don't know Mum, I'll have to check tomorrow."

I know that Glynn has always had a sharp eye for detail; therefore this was clearly not a lack of observation on his part, but something quite different. Every time I hear my proud wife recount this special story, it fills me with hope for the future. The thing is that Glynn hasn't really changed that much: I'm sure that he would be able to tell you if a new friend he had just made had the same colour skin as him or if it was different, but it would have as little meaning to this exceptionally uncomplicated young man as would the colour of the person's eyes. He has remained untouched by the world around him with regards to any prejudice of that nature.

So, what was different for Glynn? No doubt he is an 'Indigo Child' and displays many of the traits that identifies him as such. Sadly though, some of these qualities have hampered his path in life because they remain too 'unique' for the complicated material world that he currently inhabits. I do sometimes envy the down-to-earth world that he perceives, but I sympathize with the effect of the *speed bumps* that are constantly slowing down his journey.

"It's a dangerous business, Frodo, going out your door.
You step onto the road, and if you don't keep your feet,
there's no knowing where you might be swept off to."

Bilbo Baggins

Experience: The Scar

London – 1965

On my eighth birthday the world around me was 1965. In reviewing 'stuff' that was manifesting at this time, it seems to me that some things never change when you consider the world we live in today.

In 1965 Malcolm X was *shoved-off* this mortal coil, whilst Winston Churchill and Nat King Cole *naturally faded away*. The war in Vietnam continued to rage and so did the polarities that revolved around it: 25,000 concerned citizens marched on Washington to protest, whilst the President of the United States, Lyndon B. Johnson, responded by sending 50,000 more troops – umm, interesting response! The students on American campuses made their voices heard by protesting for the first time and other young people went on to burn their draft cards and, for that brave act, were subsequently arrested by the FBI. It was a time of turbulence and unrest, but also a time of liberation and revolution: it would still be many years before we saw radical change, but what was going on in the mid 60s in the minds of our young people were the seeds that would lead to a shift in the psyche of us all.

The largest blackout in the history of the east coast of America happened: from Toronto, all the way down to New York, covering an area of 80,000 square miles, the lights went out. Nine months later many post baby boomers were born as a result of this occurrence! (Imagine what would happen if the electricity went out for an extended period?) Also, interestingly enough, just around the time of this blackout, UFOs made front-page headlines by appearing in the skies throughout the United States of America – any connection do you think? Did they overload the power plants maybe? Did the darkness help us to see them more clearly? Sometimes it works like that.

And the mini-skirt made its debut. Did this have any connection to UFOs, civil unrest, the awakening of our youth or the war in Vietnam? No, but I thought I'd just put that in.

In the neighbourhood where I lived, I came into contact and made friends with a group of local children of the same age. Our playground was a small estate of apartment blocks that ringed a very simple circular, fenced, concrete, sports area. With the restrictions of school we would normally congregate properly for assorted activities at weekends; so it was with great enthusiasm that we would gravitate back to our meeting points to catch up since last we met. It was at this time that the very beginnings of adult and cultural influences would begin to colour our reactions, responses and communications with each other. As certain individuals displayed the need to control and dominate others within the group – whilst not necessarily understanding or appreciating consciously the meaning behind their actions – it was normally me who would stand to the side confused and bewildered by their behaviour. The desire to spend time with children of my own age overrode my feelings of being uncomfortable around this type of peculiar activity, but nonetheless I was always ready to retreat to the safety of my home if it became too foreign to my way of thinking. Because of this scenario I did tend to stand out from the other children, which often caused unwanted attention on my part. To avert the unpleasant words of the dominating minority I would resort to saying things that I knew would amuse and distract – and so we had the birth of the classroom clown. Thus was established a pattern for the future that would play a detrimental part in my adult ability to communicate my real feelings for fear of exposing my differences from those around me. As an adult, this self-taught survival behaviour would hamper my ability to emotionally connect without first establishing a special type of friendship with people. Even then, because of the additional learnt behaviour of others attached to this coping technique, the humour would sometimes be tinged with barbed hooks of sarcasm that would ultimately alienate any real communication of worth.

During a weekday mid-afternoon school holiday in the summer of 1965, I was delightfully engrossed in a game that required the majority of my friends hiding in the surrounding area that encircled a large horse chestnut tree. Right next to the tree stood one solitary individual whose job it was to spot anybody who attempted to reach his location before he could call out his or her name. There was definitely a name to this pastime, but I cannot for the life of me remember what it was. No matter.

I was secreted, very furtively I believed, at the base of one of the apartment blocks in a doorway. I could see several of my companions whose inability to hide properly would no doubt eliminate them from proceedings at an early stage. I often wondered how so many of my friends were able to say and do things which were so obviously ridiculous that their hope of achieving their objective was virtually non-existent. Was I the only one who ever really saw how things were?

I had been accused of cheating when playing this game in the past. The 'protector of the tree' would always be shocked to find me suddenly by his side, without any indication of how I had made my way to the tree without being spotted. I always found it quite easy to somehow distract the person by just thinking that he should look the other way, whilst I just walked stealthily, but calmly, up to his side.

Whilst I stood in the doorway about to start thinking about getting to the tree, I became aware of a bright light coming from above the apartment block. Stepping out a little from my hiding place, I looked up to see what I thought was the sun shining blindingly in my eyes. As I started to raise my hands up to my face to shade my eyes, I felt an unexpected heat burning my face. Recoiling, I stumbled back into the doorway and slumped down. As I opened my eyes again, everything was black; I just couldn't see anything. The roughness of the concrete beneath my hands as I had fallen down was now replaced by a smooth, cool surface. Although apparently in a sitting position, my feet now appeared to be off the floor and dangling in space. I was sitting on something. Attempting to focus in the darkness, I opened my eyes as wide as possible. It was only then that I realised that I was definitely no longer in the doorway at the bottom of the apartment block.

There were figures around me moving in the semi-darkness. I thought it might be the other children, but their outline was somehow different. As one of the figures moved closer to me I became totally and fully convinced that it was not one of my friends. Reaching out and holding on to one of my hands stood a most peculiar looking 'child'. It somehow reminded me of my 'friends from the hospital', but it was also quite altered at the same time. Its other hand reached up and gently guided me backwards, laying me down on the platform that I was obviously resting on. The surface felt hard. I nonchalantly drummed my fingernails against it. I wasn't really scared yet, maybe a little apprehensive, but not scared. There was something about my circumstances that seemed familiar, not least the 'child' that was by my side. The 'child' then placed a hand up to my face and rested it palm down on my forehead, it moved gently down over my eyes, closing them in the process. I felt something being pushed against my forehead; it felt a little uncomfortable and I pushed back from the pressure.

I instantly began to dream, or so it seemed. In my dream there was an angel. Her skin was dramatically white and had the quality of moonlight on virgin snow. Even though it was so pale, it did not look sickly in any way. In fact there was a peculiar luminescence to it. Her hair was long and platinum blonde in colour and hung beguilingly and protectively around her face. Her eyes were a brilliant and frosty pale blue. She was cradling me in her arms. A reassuring sense of safety and security enveloped my being as she pulled me close to her in a gentle rocking motion. She slowly lowered her incredibly stunning face and touched her pursed lips softly to my cheek. I felt the warmth of her touch and breath upon my face. Just as she broke away she whispered something that I did not understand – it wasn't that the volume was insufficient, just that I did not recognise the words.

The angel's face began to fade. It was gradually replaced by the familiar vista of the horse chestnut tree in the distance. I appeared to be standing at the bottom of the apartment block once more, several yards from the doorway that I had previously been hiding in. There was no sign of the 'tree protector' or any of my other playmates in fact. Looking up I could see that the sky was no longer bright; it was clearly much darker than I thought it should be. Now I was scared. Scared because I couldn't explain what had happened and scared because I knew I wasn't supposed to still be out when it got dark. Taking off at a fast pace, I raced across the grass, past the tree and quickly rounded the bend that led to my apartment block. Standing in front of the door to my home, I was reluctant to announce my arrival for fear of how my tardiness would be greeted. I didn't have to wait long though as the door was opened and my mother stood glaring at me in the doorway. Grabbing me by the collar of my shirt and roughly dragging me inside, she yelled, "What time do you call this? It's nearly eight thirty and you should have been in hours ago". I looked sheepishly up, desperately searching for words to explain my predicament. The words however did not come.

"Get straight up to bed young man, you are in so much trouble."

As I lay in bed, gently sobbing and slowly releasing the tension that had built up inside me, the strange events of the afternoon began to replay in my head. It had been the afternoon when I had begun the game with my friends, and then all of a sudden, it had been the evening. What about the odd looking people I had seen in my 'dream'? What had been pressed against my forehead – which was now aching quite a bit? As the tears ran down my cheeks, I closed my eyes and saw, once again, the face of the 'angel'. I remembered her comforting touch and the way she had gently rocked me in her arms. A great sense of calmness overcame me and gradually the tears stopped flowing. Presently I fell into a deep reassuring sleep. Many years later at the age of thirty, as my hairline began to recede, a small white scar could be seen on my forehead, exactly in the same place where I had been 'pressed' by the strange looking 'child' all those years before. It had been hidden under the hair growth for all that time, but now it was exposed and initially caused puzzlement until the original incident was recalled with clarity. As I looked in the mirror, I put pressure on the scar with my fingertip; it felt hard, as if there was something underneath the raised area of skin. By pressing it, I caused a most peculiar sensation: although I was applying pressure to my forehead, the sensation of touch was experienced in a line all the way to the back of my head.

The memory of this bizarre happening had always remained there, just like numerous other peculiar events that had occurred throughout my childhood, but I had never had the mental ability to process the information and evaluate the facts on any other level than the one that I had first recorded it on. One fact that I had forgotten, but was reminded of by my mother, was that when I appeared on the doorstep to my home, my head was bleeding quite profusely and had badly stained the check shirt that I was wearing. You might ask why this alarming feature was not investigated at the time – which it was not by the way – but I suspect it was reflective of the knee-jerk response of my mother that I had been 'naughty' once more; therefore "I only had myself to blame". Sadly, on most occasions, my mother's response overrode anything my Dad might have to say. I can't imagine sending my injured son to bed, head bleeding from an unknown injury and clearly confused from an event that left him bewildered and frightened, it just would not happen!

Anyway, in my opinion, I would have to say that this was most definitely some sort of ET contact. Having said that, although many of the features were similar enough to warrant this identification, there is the matter of the 'angel' that makes it stand alone as apposed to other contact events. The act of instantly transporting from one location to another without recollection of detail, had happened on other occasions; the same is true of the shadowy, silhouetted figures that appear like children, but act with a more mature nature; and finally, the ability of the said 'children' to bring about some sort of changed consciousness purely through the act of touch has featured more than once. But coming into contact and interacting with an apparently 'normal' looking human person was a new feature of my ongoing contact experiences (as far as I could remember that is).

Clearly it is worth considering that what I encountered, the 'angel' element, could have been some sort of artificial image projected into my mind by the ET intelligence, now commonly referred to as a 'screen memory'*, for what reason we can only speculate upon. Maybe she was an ET that had been made to appear differently to me. Or this angelic being might even have been a real flesh and blood ET, just not the same as the ones I normally encountered – possibly one of the types that normally 'hide behind the scenery'.

As for the scar, once more I can only speculate that either a 'scanning procedure' was carried out or even that a 'device' of some sort was inserted under the skin or possibly further into the brain itself (the pineal gland is a suggested location).

I can only say that my encounter with the 'angel' was really quite fantastic and continues to generate a feeling of wonder when recalled. I do question on what level this feeling is occurring though – for was this meeting on a physical or metaphysical level? There is no doubt that the trauma to my head was very physical, but was the incident a multi-levelled experience? And what did actually happen whilst she held me – so lovingly – in her cradling arms? What did it mean? What does it mean to me now?

"There is always one moment in childhood
when the door opens and lets the future in".

Deepak Chopra

The End of Childhood

South London – December 1969

In December of 1969, just before the Christmas holidays, my childhood came to an end. I was 12 years old. Without putting up any sort of a fight, not that I could have, I was dragged silently and painfully into the world of grown-ups. Overnight I took on board practical and emotional responsibilities that up until a couple of days before were still years away.

Because of a solitary, uncaring and thoughtless action carried out by my mother the very fabric of our family life was ripped apart.

I shan't go into the finer details of the event and the nuclear fall-out that followed, but the one emotional black eye that stayed with me for many, many years was my inability to trust. I had a problem with trust in any sort of relationship, platonic or romantic. I became a very confused boy, a tortured teenager and then an anxious and worried adult.

These disturbed emotions were reflected in every aspect of my life, none greater that in my contact with the ETs. Overnight I became scared of them and what might happen when they came. *They* didn't change; it was only how I perceived their contact that did.

As way of a survival technique, I began to bury things. I started to hide things away. I even went as far as denying things to myself.

Part Two

The Teenage Years

*"If you can keep your head when all about you are losing theirs
and blaming it on you.
If you can trust yourself when all men doubt you,
but make allowance for their doubting too;
yours is the Earth and everything that's in it.
And – which is more – you'll be a man, my son!"*

Rudyard Kipling

Experience: The David Bowie Night

London suburbs – 17th January 1973

When it comes to my recollection of 1973, the truly meaningful events of the year beyond the limited range of my own life shamefully passed me by with little or no impact at all. Therefore, as part of this project of documenting my contact experiences I am taking the opportunity to research the relevant time periods in closer detail, which is becoming an educational process on its own, for it is now clear to me that previously I was somewhat distracted.

So, here we go – pay attention – all of this will be in your final exam!

In January of 1973 an agreement was reached between opposing forces, and U.S. forces were finally withdrawn from Vietnam and American prisoners of war were released. A peace pact was signed in Paris to denote the official end to 'offensive military action'. The troops began to be withdrawn and apparently by the 31st of December of 1973 there were only 50 military personnel left in Vietnam – yeah, right!!!

I do remember the movies I saw that year: *The Way We Were, American Graffiti, Last Tango In Paris, The Sting* and, it really goes without saying, *The Exorcist*. I can also recall strutting my funky stuff to Stevie Wonder's *Superstition*, smooching on the dance floor to *Me and Mrs Jones* and stumbling drunkenly along the street in December singing Slade's number one UK hit, *Merry Christmas Everybody*.

I suppose my inability to acknowledge what was going on in the bigger world was reflective of my age at the time, which has a tendency to make a young person's view of life painfully and debilitating insular. Therefore, when I look back to this specific period, all I can remember is what happened to me.

I do not recall reading about or even being aware of the main headlines: the inhumane pantomime that was Vietnam winding down; Watergate coming to the boil; Nixon's inauguration – just in time for him to resign; there was continuing conflict in the Middle East; Picasso, Betty Grable and Lyndon B. Johnson died; Elvis gave Pricilla $750,000 in a divorce settlement; and people began to keep rocks as pets – yes, you read that right!

As far as I was concerned, 1973 had one focus: apart from the allure of the opposite sex, I was interested in one thing and one thing only – the music of David Bowie. I had heard whispers about this rather peculiar looking fellow since 1971 when his album *Hunky Dory* was released, but was unable – because of financial restrictions – to purchase it. In fact, I would not properly discover David's back catalogue until after his landmark release of 1972, *The Rise and Fall of Ziggy Stardust and The Spiders From Mars*, and that would be true for the majority of all of us.

I have so many memories from my teenage years that have a David Bowie soundtrack to accompany them, but I would like to share just one that remains especially poignant. In those days, before the high street was invaded by chain stores, there were still small, quaint, privately owned establishments of character. Although I am unable to recall the name of this record store, I can still remember how it looked in great detail. It was seductively dark, with every nook and cranny possessing a potential item of wonder. Posters of current rock stars were plastered all over the walls and the vinyl racks containing a select grouping of musical acts were lined up and down in regimental rows. A normally unrecognisable, but alluring sound blasted from speakers hanging from the ceiling, whilst the air was filled with exotic scents. The 1960s had closed its doors a couple of years before but the proprietor of this magical establishment had not been informed, and it was all the better for it, I might tell you.

Spending many hours trawling through the alphabetically-arranged album sleeves, I discovered a plethora of artists that had previously been lost in a world unreachable to my limited financial perspective. Monikers of creative wonder leapt out of the racks on album covers of multi-coloured splendour. I truly was in seventh heaven and the memory still sends shivers up and down my spine.

I had already bought the *Ziggy Stardust* album the year before, but I was now in a position to consider purchasing what I had missed. And I was going to start with – I was advised – the extremely accessible *Hunky Dory*. One feature that made the record store so much more attractive than the record department of W.H. Smith was their basement, which contained intimate, dimly lit, booths, where one could take a pack of sandwiches and spend one's lunch break listening to a new album on huge padded headphones. This was obviously intended to help one decide upon the merits of an album before purchase – and this did often occur – but for me it became a magical hole to disappear down half way through the day. I can recall with perfect clarity donning the all-encompassing black plastic headphones and settling down for my introduction into the world of *Hunky Dory*.

I had never listened to any form of music through the intimacy of stereo headphones before, so when the needle nestled into the vinyl groove and the first few bars of *Changes* were emitted, my actual world was forever changed! As each song played, the crispness and clarity of the voices and instruments filled me from head to toe with a passion that I had never known before. As the final hypnotic, swirling sound of *The Bewlay Brothers* faded into the ether, I sat there in the half-light and tears ran down my cheeks. I cobbled together the few quid for the album and wandered out into the daylight. I had forgotten to eat my lunch.

In January of 1973 I was nearly sixteen years of age. Due to circumstances beyond my control, the need to leave senior school and seek full-time employment had arisen. The small financial contribution

I then made to the home coffers was just enough to maintain the fragile economic stability for the family unit of my father, two young sisters and me. With my mother abandoning the home a few years before, the need to play a part in paying the bills had not come too soon. Although I was very intelligent, my academic career had hit the skids a while before. Some observers would speculate that it was due to the break up of my parent's marriage that my grades had dropped, but in a way I was happy to be free of the constraints of formal education. But then again, this reasoning was not to be trusted; it was, after all, coming from a fifteen-year-old boy. A tough lesson for anybody, especially an emotionally ill-equipped young person forced out into the real world before his time.

But life went on. Surrounded by other spirited individuals of a like mind, my working and social environment in the suburban town of Croydon, just outside London, kept me buoyant through this early transitional period. Distracted by a steady flow of romantic liaisons and the ability to become inebriated after the consumption of a tiny amount of alcohol, I didn't really notice that I was funding my lifestyle on a relative pittance of disposable income. I was a 'Young Turk' after all and the real sweet delicacies of life appeared to be of an intangible nature. Looking back from the viewpoint of an adult, I would wonder how on earth I was able to do so much with so little money to fund my lifestyle. I recalled dancing to the early hours of the morning in a local discothèque (slipping past the bouncers free of charge and making a solitary drink last forever), then walking miles to return home after missing the last bus; visiting the cinema to see the latest blockbusters (being let in the backdoor by one of my 'paying friends'); spending hours in the pub after work – chatting garbage, posing and romancing – and finding moments of newly discovered sexual bliss in the arms of an array of teenage beauties. It appeared to me that the 'good life' could be achieved at very little expense. It wouldn't be too long, however, before I discovered that the price to pay for some things was not always monetary.

A shared passion amongst my friends was the music of David Bowie. I had grown my hair long and had had it cut in the fashion of Bowie's alter ego 'Ziggy Stardust'; it was dyed a dark shade of red, courtesy of a bag of henna purchased from the Biba store in London.

A single, long, dangly, earring hung from one earlobe; baggy style 'Oxford Bag' trousers swung from my waist; and a tight fitting multi-coloured top completed the overall 'Glam Rock' look. I was known amongst the teenage community of the town as the 'Croydon David Bowie'; something I was very proud of as I strutted my way along the street like an exotic glittery peacock.

A new single by David Bowie was being released called *Drive In Saturday*, from his forthcoming album *Aladdin Sane*, which was going to be performed for the first time on a chat show programme on television called *The Russell Harty Show*. After discussing the up and coming event, my friends and I decided it would be a wonderful opportunity to arrange a gathering to watch the performance together. As luck would have it, the family home of one of my friends, John, would be free of adult interference on that particular evening with his parents away visiting friends. So, on a freezing cold night in mid January, John, myself and four other friends, Paul and his brother Jimmy, Sean and Mark, congregated in the living room of a third floor apartment, just a few miles from Croydon. The air was charged with excitement as this small group of teenage boys sat about in anticipation of the night's event. The music of David Bowie blared out from the stereo, which only added to the combined energy and suitably prepared the expectations of the group. Although I was not drinking on that evening, because I was half way through a course of antibiotics, the rest of the boys were making their way through a few six packs of beer and thanks to the excess donated by my abstinence, some were drinking more than others. By the end of the evening, Paul was unable to stand up properly and, after ordering a cab, his sibling helped him downstairs and the two left for their journey home a few miles across town. After bidding farewell to Sean and Mark, John and myself began to clear up the mess from the evening's jollities. As I lived in South London, many miles away from my current location, John had previously invited me to stay over.

I was shown to the bedroom left vacant by John's young sister, who had gone away with her parents. After relocating a colourful selection of soft toys from the bed, I removed my clothes and slipped naked under the freezing cold bed sheets. Due to the reduced length of the child's bed, I curled up my legs and lay there staring up at the large picture window that spanned the complete width of one of the walls. There were no curtains and the limited amount of illumination from the streetlights cast faint shadows in the room.

I felt restless, unable to sleep in this unfamiliar and cramped bed. From my position I had an unrestricted view of the cloudless night sky and watched as a patch of stars twitched and twinkled in the velvet blackness. As minute after minute passed, I began to feel more and more uneasy, aware that it wasn't just the present sleeping arrangement affecting me. Just out of reach was an inkling of something else, something that compounded this growing paranoia with each passing moment.

And then came that strange smell. It reminded me of Christmas time somehow. Its perfume was sweet. Almonds? It really became quite sickly and overpowering. My head began to swim as if my senses were intoxicated. Like a current of warm water lapping at my toes, numbness began to rise through my feet and into my legs, eventually saturating my whole body. Although this frozen state had elements of familiarity, I was beginning to experience a growing sense of anxiety, because I knew what was coming next. In sheer desperation I was still able to screw my eyelids closed in a vain attempt to avoid the anticipated next stage.

I could hear unusual noises in the room. It sounded as if something was scratching against the linoleum floor. Did John own a dog? Not in an apartment – surely not. And then there was something close to my face. Even though my eyes were still firmly closed, I could feel the presence of something. Although the last thing I wanted to do was actually open my eyes, there was a force present in my mind compelling me to open them. Within an inch of my face, there was something looming over me. I had to look. I had to open my eyes to see. It was another face. A pair of black eyes fixed on my own with an irresistible hold – a kinetic energy enfolding my conscious thoughts as if they were not my own. An exotically egg-shaped head hovered above me, and although somehow familiar in appearance, it brought little comfort as I continued to experience an invisible probing force invading my thoughts. And then, just as quickly as the high strangeness had taken hold, a sense of calm began to take over. A voice in my head assured me that "everything is all right".

The figure stood upright and stepped back to reveal a host of similar creatures standing motionless all around us. All at once, I felt immediately safe, intuitively knowing that, in truth, no harm would come to me and that everything was all right.

Slowly, very slowly, like sand trickling from an upturned egg timer, feeling began to return to my body. I felt a long, slender hand reach down delicately beneath my back and helped me rise up off the bed. A pair of hands now reached out and swung my legs adroitly to one side with my feet skimming the floor. Hands continued to appear in support and then I was standing on wobbly legs as the feeling continued to return.

I stood stock-still as two figures stepped forward and with assured pressure placed a hand on each of my elbows and began to walk me away from the bed towards the window. And once more, I knew what was going to happen next, and although I tensed with apprehension, I was not fearful; for whilst the circumstances remained surreal, the proceedings were familiar.

As we came within a couple of feet of the window, I instinctively closed my eyes, remembering the peculiar sensation I would now experience. My two guides edged me forward with a confident step and then I knew, without actually seeing, that we were now passing through the solid wall and window, completely unhindered by its solidity. A sensation of stickiness wrapped around my naked body as if I were walking through a fine net curtain. As I emerged from the other side of the window, I immediately felt my feet leave the ground, the pressure from the guiding hands on my elbows increase and I appeared to rise swiftly. I shivered as my exposed body experienced the penetrating coldness of the night air. Feeling the sudden change of temperature, I opened my eyes, only to have my vision instantly blurred by a flare of bright light that hung suspended in the sky above me. I could still feel the firm grip of my companions at each elbow, as they remained apparently fixed by my side as we rose towards the light. Responding once more to an urge to close my eyes, I felt a most peculiar sensation: as if I were no longer there, as if the physical universe had somehow disappeared and I was just hanging in a void of nothingness.

For all I knew a million years could have passed as I attempted to reach out with my mind to connect with something solid. The sense of oblivion flooded each thought and continued to do so, for how long I will never know. And then it was gone. I opened my eyes and realised that my sense of being somewhere had returned.

I lay flat on my back, still completely naked, upon a cold hard surface, staring up at a dimly lit ceiling. Instantly aware of movement in my gloomy surroundings, I swivelled my head to one side and saw recognisable outlines moving backwards and forwards. The small figures that I had just previously encountered came in and out of focus, as they appeared to be hurrying to and fro. Memories flooded my mind as I recalled a plethora of familiar, yet puzzling memories; memories that I knew were always just below the level of my daily conscious awareness; memories that continued to nudge at my mind, continuing to colour, influence and manipulate my take on the world and impact upon my belief systems. Why did their accessibility fluctuate? What were the controlling factors of this recall? Who governed and manipulated the process?

Literally within the blink of an eye, I was surrounded by a swarm of these odd looking little fellows. Each one took turns at moving within an inch or two of my face, forcing me to slightly recoil as they pressed forward.

Eventually this unexplainable behaviour stopped and I was manoeuvred by unseen hands to an upright position with my legs turned to one side and I stepped down onto the floor. Opening my eyes just enough to see in front of me, hands were once more placed at my elbows and I was walked hesitantly out of the room I found myself in. I saw that I was in a hazily-lit corridor that curved away to my right. On the wall, just a couple of steps away was a bench. It had no obvious support, but appeared integral to the wall, as if it had melted, creating a protruding lip only about 24 inches off the floor. My rigid frame was turned and let down onto the low-lying seat. Shadowy figures appeared to back away from me and returned through the opening that we had just come from.

A surge of vulnerability came over me, bringing me instantly out of my shocked and dreamlike state and my sense of nakedness was overwhelming. Like rising up through the depths of the ocean, my consciousness broke the surface as I caught my breath and filled my lungs with the reality of the situation. I felt alone and susceptible. There was no discernible sound. No recognisable light source. Nothing familiar enough to make me feel safe. There was no sense of anything; everything felt completely alien. Although I was breathing normally, I felt as if I were in a vacuum – yet unable to really comprehend the true meaning, as if I was somehow caught in a moment of time.

As I began to look around me, I remembered where I was. I had been here before. This had happened many times, but it always took me a while to access those memories. Why was that, I asked myself? Why did it always come as such a shock to my system? The walls curved away from me on both sides and receded quickly. It was as if I were sitting in a pocket of light, whilst all around remained dimly lit and still with no apparent light source. Turning my head to one side, so I faced the wall behind me, it felt as if the ambient light was somehow emanating from the wall itself. Moving my eyes closer still, I could see hundreds and hundreds of tiny pricks of light breaking through the make-up of the wall's surface. But as I moved my head back just an inch or two they disappeared. This moment would be brought back to me at a later date when watching the fragile strands of a fibre optical lamp as its lighted points swayed hypnotically before me. Not the same as the wall at all, but similar.

There was movement from the corner of my eye. I saw silhouettes staggering towards me with an unruly and uncertain movement. For the first time I was able to *put a name to a face* as the expression goes. There were two small *people* that I was now able to distinguish as 'Greys': tiny, fragile figures that stood no more than four feet tall, bulbous headed extraterrestrials with dark penetrating eyes that remained fixed on me throughout, while their spindly, disproportionately long arms appeared to be supporting somebody else between them in an obviously shaky, unnerving manner; it was a bulkier, taller more recognizable human being who gently wobbled as if unable to stand under his own volition without the support of his alien hosts. It was Paul. *It was Paul*, my inebriated companion from earlier on in the evening, clearly still drunk and being held up by two small alien beings. Where's your camera when you need it?

Paul's eyes were closed and his chin hung droopily on his chest. It was clear to me that without the support of his two minders he would probably lose his balance and fall flat on his face. With enormous effort, he lifted his forlorn head up and whimpered pathetically. It's totally crazy what inappropriate thoughts go through our head at times, but at this surreal moment this is what I came up with: 'Paul would not be too happy with himself when he eventually sobered up, for he'd been explosively sick down his newly acquired David Bowie t-shirt!'

I stood up and moved towards them with a determined stride. I was quite surprised at my action, considering the circumstances and, based on their response, Paul's two supporters did not expect it at all. They instantly released their grip on his elbows and took one step each to the side and backwards. Motivated by compassion for my friend and not considering anything else, I leapt forward to catch Paul as he stumbled forward. Embracing and pulling him to my chest, his head slumped heavily on my shoulder. I whispered into his ear, "It's all right now mate; I'll take care of you."

Even with my own mind-boggling quandary tugging at my reasoning for explanation, I was still able to reach into my own humanity and find the strength of will to help my friend. Although it would be many years before I would be able to acknowledge the compassion and courage of that moment, the memory would eventually return and with it would come a much-needed sense of empowerment as a human being. I am not taking praise as an individual, but I would like it to reflect upon what human beings as a whole are capable of when the chips are down. There is something beautiful that resides in the heart of humankind, something that makes us so very special and it breaks my heart when I see behaviour that comes from a much darker, unacceptable place, which puts us all to shame.

Just like any normal adult human being, the two small aliens had clearly found it extremely hard to cope with a drunken teenager! Whoever they are, whatever they are, they had taken on board (in more ways than one) a little more than they could handle. As for me, the need to secure and make safe my inebriated friend was not something new — after all, I am human and 'they' were witnessing human behaviour. I would later reflect on how capably I took control of the situation, even though the surreal circumstances could have easily overwhelmed my capacity to do so. On this and many other occasions that were to follow, I would surprise myself in my own ability to rise above my circumstances. It would always appear that the more demanding the state of affairs, the more resourceful I was able to be. And isn't that true for all of us? Isn't that one of the most wonderful traits of being human?

As I propped Paul up, just content to maintain a vertical position, I felt long, familiar fingers grasp my elbows and begin to once again guide my movement forward. As I coaxed my friend's feet along, whilst all the time reassuring him of his own safety, I watched our custodians with a wary eye. I could sense an uneasiness coming from them, as if for once they were not in control of the situation. It felt to me as if they had relinquished just a little of the control they normally held. Was this to secure a healthy completion to their operation, one that had clearly not gone to plan? All of these thoughts and feelings were obviously intuitive, as no outward sign of our keeper's emotions or thoughts were ever apparent.

Initially stepping out of the pocket of light, it did then seem to follow us like a spotlight as we moved along. As we turned the corner of the corridor, I once again experienced the stickiness of moving through some sort of invisible energy field, as I had previously felt when I was whisked away from the security of John's apartment. Without apparently breaking step, my vision momentarily blurred, only to be replaced a second later by the sight of an unmade bed with crumpled sheets lying in a heap on the floor. This was not the bedroom I had left earlier that evening, but a normal looking one nonetheless. I immediately *understood* that this was in fact Paul's bedroom. I positioned my friend on the edge of the bed and swung his legs up, whilst also attempting to support his head as it fell heavily to the pillow. I turned him on to his side for safety. Paul groaned about feeling sick as I covered him with a sheet and blanket. Looking around his room, I saw a waste paper bin beside the door and retrieved it so he could throw up into it if required. It was only then that I realised that the ETs were no longer at my elbows, but I knew without looking for confirmation that they were there in the room observing my every move.

Straightening up from my crouched stance, I heard Paul's breathing become more rhythmic and a low snoring began to bubble from his lips. Feeling secure that my friend was now safe, I turned to face my watchers and as I did so the room appeared to go completely black. The ETs appeared to instantly shrink in size as if they were moving quickly away from me or I from them. The darkness enveloped everything; nothing was visible at all, not even the hand in front of my face. The sense of nothingness returned; of simply being nowhere. Then I felt as if I was lying down on something soft and springy; of being wrapped with a warm and comforting cover; and of feeling long, slender, delicate fingers tracing a repeated line across my brow. That was my last conscious thought until the morning. In fact, when I did finally wake up, the memory of the previous night's strange experience would be temporarily inaccessible. Even more peculiar to me was that I did not wake up back in my friend John's apartment.

As I roused the next morning, I had no immediate recall of the events of the night before, except that peculiar sensation one experiences after a late night of partying and the reluctant and uncomfortable feeling of "What did I do last night? I hope I didn't do anything embarrassing!" There was certainly an inkling that something had occurred of a dubious nature, but nothing more. In fact, as I lay there stretching and assessing my body, apart from this lurking suspicion, I couldn't help but come to the conclusion that I must have had a refreshingly uninterrupted night's slumber.

The room was still extremely dark. The illuminated face of my watch read 8.45am and, although we were still in the depths of winter, I knew for sure that it must be light by now. Squinting in the darkness, my eyes became more acclimatised to the lack of light and the first thing I noticed was the unfamiliarity of my surroundings. Reminding myself that the curtainless window should be allowing more than enough light in, I instantly saw that there appeared to be only four blank walls surrounding me. The tiny child's bed that had previously been tucked against a flat wall was now placed in the middle of the room. And as for the reduced dimensions of the said bed, I could now stretch out my legs without encumbrance. Continuing to scan in the limited light, I could see several stacked cardboard boxes at the foot of the bed and numerous unidentifiable objects placed here and there against the walls. Turning round in my horizontal position, I viewed a rough wooden staircase leading up towards a door that stood slightly ajar with a slit of light creeping down the steps into the room.

Deciding to investigate my bizarre predicament, I sprung from the bed and looked for my previous night's attire. The neatly folded clothes that I had placed on the end of the bed were no longer there. Tugging the brightly patterned bed cover, I wrapped it around my naked, shivering body and ascended what appeared to be my only way out. Apprehensively standing before the door at the top, I could hear the familiar and reassuring sound of crockery being stacked. Stepping through and into the room beyond, I encountered the welcoming environment of a large family kitchen, and there, in the far left-hand corner, was a red-haired woman in a fluffy white towelling bathrobe, stooping to retrieve plates from a dishwasher and placing them on the side of the counter. Although it would be safe to say that I was in a state of mild confusion and bewilderment, I was still able to summon up a convivial greeting to announce my presence.

"Good morning," I chirped, ever the polite young boy.

The response I received was unexpected to say the least. I would have accepted the possibility that I might have made her jump and that maybe a plate might have slipped from her fingers. I knew without any doubt at all that this woman must have, at some time in her life, encountered another fellow human being; therefore her actual response to my greeting only added to my overall bewilderment of the situation that had presented itself thus far.

It was abundantly clear that the woman did not know who this semi-naked teenager was or what I could have possibly been doing in her home. By her animated response, I had no doubt about that. Amidst shrieks of apparent terror on her part, she let forth with a tirade of colourful Anglo-Saxon expletives that convinced me of all of the above in a matter of seconds. I would imagine that my face must have reflected a range of contortions to match her outburst. Raising my voice to the required volume in order to be heard, I attempted to convey the relevant information to hopefully explain my current circumstances and possibly placate this emotionally overcharged screaming banshee. I did not believe that she could possibly have heard me at first, but I continued on regardless: my name; that I was John's best friend; and I had been invited to stay over. As the red-haired woman appeared to stop ranting long enough to take breath, I then took the opportunity to give it another shot at a slightly higher volume. And it seemed to do the trick.

Although she had stopped yelling at me now, her twitching face showed that she remained in some sort of panic mode. It was more than apparent that her brain was now in overload as it attempted to process my information and reach some sort of informed understanding of what was turning out to be her worst nightmare. I was surprised that she managed anything looking at her face, which was turning all shades of crimson.

"Who... *are you*? Who... is John? *What*... are you doing in *my* house?"

And there it was: succinct, clear and precise.

I took a breath, tried to answer, but found no way of responding. I knew all the answers, but appeared to be physically incapable of expressing them. I ran her questions through my mind once more and then squared them up with my understanding of the situation. To enable me to answer her questions, my answers should have seamlessly dovetailed with them, but everything was more than just a little askew. Ridiculously, and inappropriately as it might sound, my next response was one of laughter – call it gallows humour if you will – and it was instantly clear to me that I had made the incorrect choice. I believe now that the only thing that stopped the woman snatching a carving knife from the dishwasher and hurtling it ferociously at me was the arrival of a third party into the arena.

I felt somebody behind me and turned to see the bleary, red-veined eyes of my intoxicated companion from the previous evening's jollities – it was Paul (although at this stage I had no recall of the high jinks that had occurred after the social gathering). A weary, half-hearted smile creaked from his dry lips, as he croaked hoarsely, "Hey man, how's it going?"

From the heated exchange that then occurred between Paul and the red-haired woman, it was apparent that the relationship was, without doubt, one of mother and son. It was also apparently clear from their dialogue that somehow I was in *their* home and not

John's.

Paul's mother – she of the red hair and a healthy inspired range of quaint parlance – strode across the kitchen with a determined force until she was face to face with her, now cringing, offspring. As best as I could make out, as I was now attempting to assume the mode of invisibility, she expressed her displeasure with finding uninvited, half naked, obviously retarded, friends of his appearing in her kitchen like a genie out of a magic lamp. Paul rubbed one tired eye and then the other, all the time looking more and more confused, which resulted in both of his eyes appearing just that little bit more bloodshot and tired. In the blink of an eye a hand was raised and sent crashing down across the side of Paul's head, rearranging his already unkempt hair and, I could only imagine, adding to his pounding hungover state of being. She then turned towards me. Without even thinking, but responding with a well known knee-jerk survival technique, I ducked.

"You'll get yours young man, but I'll have to know your name first."

She shoved past us both and left the room.

Her continued rant could be heard from the hallway as the kitchen door continued to swing to and fro on its rising hinges. Paul and I stood there looking at each other. As the sound of his mother finally faded away completely, the silence in the kitchen was deafening. In those first few moments a million and one thoughts must have been processed, but not a word was said. I struggled with the pressure of the real truth as it attempted to barge past whatever had been placed in its way, but to no avail. Whatever security door had been fitted between the ET contact event of last night and the cold reality of this morning, nothing was going to budge at this stage. I can only imagine that what was going on for me was also being experienced by Paul. For the same reasons or for different ones, I would never know.

"I thought you stayed at John's last night?"

"I did."

"So what are you doing here?"

"I don't know."

Okay, we had got that out of the way. Now where do we take it from there?

Silence. Unheard sparks inside each of our brains. But still silence as words would not form. Then we were joined by a third party and it was not Paul's mother – thank god!

"I thought you stayed at John's last night?"

"I did."

"So what are you doing here?"

"I don't know."

It was James, Paul's brother.

Then, both Paul and James asked the same questions again, phrased and worded slightly different, but this time charged with a clear hostility on their part. The questions continued on a round-robin, the volume increased and I now felt intimidated and threatened. That was until I raised my hands in a gesture of surrender: "Enough, enough, *enough*. I do not have an answer for you guys; I am just as confused as you. But let's get something clear here and now, shall we?" I waited for their agreement and then continued. "I did not leave John's home last night, just after you left or at any other time. I did not walk across town, without knowing where you both lived – I've never been here before remember – and I did not break into your house and slip into bed in your guest bed in the basement."

No response from the brothers.

"*And*," I added with extra emphasis, "*I did not do any of this stark-bollock naked!*"

"But you're here," they bellowed simultaneously.

"I know," I replied, just as loudly. And then, more softly now, but just as passionately, "I know… *I know*."

We all stared at each other. There were no words. There were just thoughts; thoughts that continued to run slap-bang into the wall of reason that stood before us.

"Phone John," Paul said to his brother.

John was contacted and confirmed that his guest wasn't there, but a neatly folded pile of clothes were on the end of his sister's bed, where John had last seen me. Within a short while, John had arrived and I was to regain a little dignity by being able to dress. We all sat down and shared our memories of the events from the night before.

Everything matched, except where certain fragments were lost for Paul due to his inebriation, but the main points of the evening were all the same. The boys told how they went home and how James had put his brother to bed. John spoke about how he and I had cleared up the pizza boxes and the empty beer cans and then we both retired. He added that when he had left to come over this morning that the front door had been locked and bolted from the inside and he also pointed out that his home is three storeys high. My contribution was to say that I had found the bed very cramped and cold and that I had gone straight off to sleep. Because our other guests of the night – Sean and Mark – had not been directly touched by the fall out of events, we decided not to discuss the matter with them at that time. However, we did bring up the topic with them at a later date and their recall of the evening's proceedings did match ours up until their point of leaving.

What we were being presented with was totally impossible and the more we spoke of the details the more we underlined the absurdity of it all: I was *here* when I should have been *there*. Paul and James' initial hostility towards me, based on their belief that I must have somehow broken into their home began to wane as we continued to explore the dead-end conclusions to the circumstances.

"We have an alarm you know," James mumbled, more to himself than to anybody else.

Paul nodded. "Dad is really anal about setting it, so there is no way Steve could have got in undetected."

As the overall anger of our group subsided, it was replaced with something else, something much more frightening. It felt like fear. We were all terrified of this new reality, a reality that had somehow leaked out of the normally safe and structured confines of reason. We were actors in an impossible scenario. There were no answers, no way out, no escape route or anywhere to hide from the facts.

We never spoke to each other about that night ever again. It was mentioned to other people outside of our tight little circle – in part that is, with the more unbelievable aspects being left out – for after a while I acquired the nickname of the 'Three-mile streaker', because that was the distance between John's apartment and the home of Paul and James. But as for the group of people that spent the evening together on that freezing cold January evening, not a word was ever mentioned. Having said that, I remembered wanting to bring up the subject on many occasions, but I knew it would be pointless. I only ever had to just think about doing so whilst in the company of the David Bowie Night co-conspirators and somehow an invisible alarm system would be activated and I would be stared down until the moment passed. After all, where on Earth would we go with it? I had already been down that cul-de-sac a thousand times in my mind. I still saw it in my mind's eye though; it remained trapped and unexplainable for a long while after the event.

I eventually lost contact with these young men as I moved from working in the provinces to central London. I often wondered though if any of my friends ever thought about what happened on that night. Did Paul's memories of the event finally return as they eventually did for me? When the barrier was finally lifted and these memories were eventually, and gratefully, released, they returned with a bright and precise clarity.

Apart from the lasting impact of what happened on that night, I am left with the sense that the 'beings' I came into contact with have remained present in my life ever since and are more like *human beings* – or visa versa – than it is generally considered. Or, might it be said, that we are more familiar with their way of thinking than we might like to believe. I feel that there is less alienation in our contact than we are truly aware of. Of course, there are possibly many theories why this might be: none less forceful and convincing than our fundamental point of origin. In the continued reports that come back via the Experiencers during their contact experiences with this remarkable intelligence, we might consider that the ETs are more *resident* than *visitor* and avoid subscribing to the tabloid depiction of them as alien in every sense of the word. Their actions and behaviour might appear peculiar at times, but then again do we not respond in a similar fashion to different cultures already present on this planet? I would bring your attention to the act of caring that the ETs showed whilst dealing with Paul's predicament and their apparent need to bring about a safe and happy conclusion to the situation. Wasn't their behaviour, at that precise moment, clearly more human than alien? If our journey of evolution has proceeded along the same path, but our Visitors just began their voyage a lot earlier than we did, then might it be possible that they acquired or developed different ways of conducting themselves along the way? Maybe there are more similarities than differences; and could that not be said of the many different and varied people that inhabit this planet – especially the ones that we are in constant conflict with, just because they look and act differently to us?

There is one last aspect of this experience that I would like to investigate. It is only now that I realise how it has been overlooked over the years during discussions and investigations relating to the details of the event. It is only now as I record the experience in this format that I am made aware of how important it was that Paul and I made contact during the event. As I continue to expand upon this thought, I'm sure you will agree with its relevance to the proceedings of that evening.

The lads that I socialised with on that night were relatively new friends; I had only known them for about a year. My more established group of friends, those from my neighbourhood and from school, were located where I still lived. This new group were all work-related people and had come in to my life over the previous 12 months.

Now, the questions I believe are important are these:

- How long had Paul's contact experiences been going on for or was this a 'one-off' occurrence?
- Would our individual contact experiences have occurred if we had not been together during the evening?
- Was it just a fluke that two independent Experiencers were taken on the same night, considering the fact that we originated from separate locations or did we have a connected ET history before we met?
- Would I have known Paul was on board if he hadn't been drunk?
- Did the ETs decide to seek my assistance in their dilemma?

I'm sure that you the reader have more questions concerning this aspect of the event. I wonder why nobody has ever highlighted this connection before? Even when Budd Hopkins* (leading UFO and Alien Abduction researcher) investigated my case, these questions were never raised. But I had never considered it, so why should anybody else. Am I putting too much importance on this? I truly believe that my contact with Paul on that night, how I responded to his dilemma and the observations made by the ETs are probably the most significant features of the whole experience. The 'humanness' of our joint behaviour is the key as far as I am concerned and deserves further thought and consideration.

"Memory is the diary that we all carry about with us."

Oscar Wilde

Food For Thought: Memory

Part 1: Access All Areas

How does the human memory work? After researching this question, one thing is clear: I don't think anybody really knows. And to their credit, on the whole, the 'experts' are surprisingly honest in admitting the fact. Obviously, they have a pretty good stab at coming up with a million and one theories, but when it comes down to it what they present is conjecture and hypothesis. And that's okay, because hopefully, these are all signposts on the way to the destination of the one truth.

For the most part, I think we all agree that memory is the mental process of retaining and recalling information and experiences. It is the method of taking events, or facts and storing them in the brain for later use. There appears to be three types of memory: sensory, short term and long term.

Sensory memories are momentary recordings of information in our sensory systems, evoked via our five basic senses: sight, smell, sound, taste and touch. Although sensory memory can be relatively brief, different sensory memories can last for different amounts of time. Iconic memory is visual sensory memory and lasts for less than a second in most cases. Echoic memory is an auditory sensory memory that normally lasts for up to four seconds. For example, if one was to inhale a certain fragrance, the olfactory tract in the nose sends signals to the brain's limbic system, which helps store the memory of the smell so that when one smells it again, it is possible to recall it.

What is sometimes referred to as the 'Working memory', the short-term memory, is the recording of information that is being used in the here and now, and which only lasts for about twenty seconds. When the brain receives signals of information, it can then be repeated over and over until it is finally stored away; this process is referred to as a 'phonological loop'. The information will be ultimately lost though, unless a repetition of it occurs within a limited amount of time.

Now we come to long-term memory, which is the capacity to deposit and hold information over an extended period of time. This can be unlimited to all extents and purposes since it could have been one minute or one year ago, and the information can still be retrievable at any time in the future. There is a school of thought that postulates that parts of the long-term memory can be permanent, while other areas may eventually deteriorate over time. But, as I have previously said, with all due respect, this is conjecture and hypothesis.

There are three main areas of long-term memory: procedural, declarative and episodic. Procedural memory includes all motor skills, which though slow to acquire are much more resistant to change or loss, such as the ability to ride a bike, for example. 'Declarative' memory is used to remember facts, such as names, dates and places. While relatively easy to learn, they can also be very easy to lose – "I bet he forgets our anniversary again this year". Episodic memory is the record of events that one stores throughout one's experience. There have been recent studies that suggest these events, as soon as they occur, are sent to a temporary part of the brain called the hippocampus, and that over time they are removed for permanent storage to an area called the neocortex.

It is believed that certain parts of the brain are used for the collection and storage of different memory types. For instance, in the hippocampus it is believed to be used to transfer memories from short-term to long-term. Additionally, it helps to store spatial memories (the recording of information about ones environment and spatial orientation) with the thalamus, which is a collection of nuclei that relays sensory information from the lower centres to the cerebral cortex. As well as the spatial memories, the thalamus can also help deposit emotional memories within an area called the amygdala. The amygdala is a nuclear formation that assists the storage of both conscious and unconscious emotions, and it too is part of the limbic system. Motor skills, the knowledge of social behaviour and the demonstration of personality are stored in the area of the brain called the prefrontal cortex. The main part of the brain that is concerned with motor co-ordination, posture and balance is called the cerebellum. To enable an overall picture of all of the areas we have spoken about, it is helpful to understand that the hippocampus, the amygdala, and the cerebellum are all part of the brain's limbic system.

I've certainly got a better picture now of how the brain is 'put together' and maybe a little wiser of how the memories *might* be processed through its construction, but I get the feeling that the 'experts' are still fumbling in the dark to a greater or lesser degree with what has to be the most complex part of the human anatomy. I do want to make it clear though, that my comments are not meant to be some sort of derogatory observation on the scientific community's possible lack of knowledge. *They know as much as they know at this moment in time.* This has always been the way, hasn't it? We can't know everything all the time. There is a progressive ladder that the human race climbs in its acquisition and understanding of knowledge and wisdom. Therefore nobody can criticize another for not having all the information, but I do wish people, especially scientists, could just openly say 'we don't know'. Can you imagine how much we didn't know in the past compared with how much we know now? Today's 'magic' is, after all, tomorrow's science.

To sum up our present understanding of how the memory works – for the time being anyway – I will say that it is the storing of information in the brain over a certain period of time. You will recall that there are three kinds of memory: sensory, short-term and long-term – all of which use different parts of the brain to store information, and which can therefore be stored for varying lengths of time.

Memory is also a relevant topic to many biological issues in the world we live in today. There are new studies being carried out to connect memory to the dream state of humans and also to the deterioration of the brain in the totally debilitating disease Alzheimer's and dementia. One thing that we do know about memories is that they are directly or indirectly connected to the brain (as a holding bank or a transfer/processing centre). It is speculated that one day we might be able to look into somebody's brain (or wherever the information is stored) and replay their memories or maybe even retrieve memories that might have been previously lost. That ability might be closer than we imagine it to be.

Up until 2001, I believe my memory worked exactly the same way as everybody else's. 'Stuff' happened to me and I recorded it. Like computing, each recording was sited in a certain file, then placed in a relevant folder, and finally stored in a specific location dependent on my need to access it in the future. Sometimes I could remember 'stuff' easily, whilst at other times it took a moment to recall or maybe not at all. Clearly that explanation is meant as an 'Idiot's Guide' to how the memory works, because I know that the true explanation is much more complex, more complex in fact than we have even begun to imagine.

Because of ET contact experiences that have occurred relatively recently, it is true that how I now record some events and then turn them into recorded memories has altered, but the most usual tool that has come out of this transformation is how I then access them. I believe that all of my contact experience memories have been placed in a special folder all of their own and I now have a direct, unhindered link to retrieving them. I am not suggesting that I have achieved this on my own; the ETs that I am in contact with must have carried out a procedure of sorts to achieve it. Before this time, when it came to contact events and the memories connected to them, my recall worked like this: sometimes, when an experience would occur, especially when I was a young child, I would have a perfect recall, but it was very simplistic and naive and not hampered by the belief systems that I would learn/adopt at a later age, systems that would corrupt my view of the world around me. At other times – and this would be more often as I went into my teens and adulthood – something would happen and a form of instant amnesia would kick-in, brought about by the ETs for reasons of their own, although, I believe, always done with benevolent motives. Recapturing these events would return via a range of memories: from a wisp of recall all the way to a full-blown remembrance of what had transpired. I sense that this was more of a 'leakage', rather than a conscious action occurring on my behalf to release the said memories. I suggest that this might be reflective of most people.

A lot of these past events would be like old wallpaper, which would hang in the corridors of my mind. As I passed through these areas on a daily basis, just like we go through the real physical locations of our homes, we know the 'wallpaper' is there, but we fail to acknowledge it on any real significant level of appreciation. So, when some of my more repressed memories of ET contact returned over the last nine years, I wasn't surprised, because on another level I had always been aware of them, but because of a 'programme' that had been in place at the time of the experience they had been disguised as mere 'wallpaper'. They had never been erased, but the blurred picture had just been brought back into focus.

Therefore, it is my belief that I have now reached a specific time in my life where it has been deemed relevant for me to have access to all past contact event memories, and for all future contact experiences to happen whilst I am fully conscious and for the record to be placed directly into the pertinent folder that reads 'Access All Areas'.

To be continued in:
  Food For Thought: Memory
  Part 2: Does Our Heart Have Memory?

"Second to the right, and straight on till morning."
That, Peter had told Wendy, was the way to the Neverland.

James M. Barrie, *Peter Pan* (Chapter 4)

Experience: Little Star and Peter Pan
The Building Bricks

Location unknown – 1975

Comparatively speaking, by the age of 18, although not totally aware and sensitive to the world around me, I was certainly better than I had been over the previous few years. Therefore, there are certainly memories from that time concerning the world I lived in that come back to me unaided and without the need to Google the headlines.

The UK had its fair share of domestic problems that year with an horrific attack carried out by the Irish Republican Army (IRA) that saw the Hilton Hotel in central London bombed killing two people and injuring 63 others. Whilst later on in the year they also detonated an explosive device outside a tube station – it was Green Park in London – which had only one fatality, but 20 casualties. During another incident on the street, a fleeing member of the IRA shot and killed a London police officer as he gave chase. But the high/low-light of the year, the one that came to be known as the Balcombe Street Siege, came at the end of the year, when a group of IRA terrorists broke into a London flat and took the residents hostage. The siege ended after six days with the gunmen giving themselves up to the police.

It certainly was a very grim year with the world of terrorism coming right to the very streets of London town. Up until that moment we had only ever seen this type of thing being played out abroad via our television sets. This was reality. I can still remember sitting in a pub one night and hearing a huge bomb being detonated somewhere in London and thinking "This is our life now I suppose".

And what was the response to these troubles by the government of Great Britain? Were they talking to the IRA about ways they could bring it all to a peaceful conclusion? Were they addressing the issues? As far as I know they were just responding to what was happening as it happened. One thing they did do constructively was end the internments of suspected terrorists in Northern Ireland – whoopee-do!!! It's good to know that the elected governments of our countries are always acting so constructively with only our interests in mind. Sarcasm! – Just in case you weren't sure.

One cosmic beam of light in a year of menacing darkness was the first episode of the groundbreaking British sitcom *Fawlty Towers*. Thank you, god(s)!

As time passed, the sharp, bright, conscious memory of my contact experiences took a step back; never totally receding as such, but just going sufficiently out of focus so as not to bring too much attention to themselves and distract me from my day-to-day life. One might wonder, due to the high strangeness of the events and their challenging nature, how this could possibly be the case. I would imagine that for most people unexplainable events are normally limited to such things as 'where have my car keys gone to?' They irritate and frustrate, but they don't have a long-lasting effect. However, human beings do appear to have the ability to compartmentalise what cannot be explained or processed. It allows us to move on, to get on with our normal lives. If we didn't have this ability, I do wonder what might happen. I believe though that there are levels of compartmentalisation. Whatever level my experiences were filed away in still allowed me to live my life, but how much did they continue to colour and influence my ongoing belief systems? If there remains some sort of influence, at what level does this occur? When we enter our home does the colour and pattern of the interior decoration affect us? It's always there and in some way or other our eyes are always taking it in, but how quickly or easily, if asked, could we describe it? I suggest that my experiences were like that: wallpaper of the mind, hanging in the corridors of my subconscious, their influence constant, yet subtle as far as my awareness is concerned. What invisible threads continued to pull upon my emotions and influence my decision-making? It is said that our brains are capable of recording every thing we have ever seen or heard; so do we remain connected to these records or are they filed away with no way of manipulating future thinking. As the colourful palate that is teenage life created new and exciting experiences for me, the surreal element of these emotive and challenging chapters from my past apparently remained – in part – inaccessible in my subconscious.

An event occurred in early 1975 that brought a new dimension to my contact experience and had a profound impact on my life, which continues to reverberate to the present day. A restless sleep pattern has always been a regular feature in my life, so on this particular night the fact that I had been tossing and turning in bed for a couple of hours came as no surprise. Although my head was filled with a hundred and one thoughts that did not aid restful sleep, I was also experiencing something new and different. As I lay there in the darkness of the room, a regular pulse that started in the muscles of my neck rippled out like a spark of electricity and travelled down the right side of the body, making my arm twitch and pulsing through my leg until the electricity seemed to be emitted from my toes with a pronounced jolt. This pattern of spastic movement would occur every few minutes on a regular basis. I could feel my fringe sticking to my forehead as if it was being stimulated by static electricity. Gradually, and moving like a gentle flow of warm water, a numbness began to creep from my toes up to the knees and eventually all the way to the top of my legs; the feeling of numbness that had been experienced on previous occasions. As this strange lack of feeling continued through the rest of my body, I became aware of movement in the room. As if rising out of the floor at the bottom of the bed I saw a familiar outline standing there. As the shape came closer, I recognised The Witch. I struggled unsuccessfully to turn my head to see her more clearly, realising that I was unable to move any part of my body except the eyes. The Witch leant over me as her eyes made contact with mine. There was no reflection in the large impenetrable black eyes that hovered above my own, but I sensed, and remembered, a profound feeling of affection and care that was emitted from this bizarre person. My initial trepidation was instantly replaced by one of well-being. I felt safe. I felt safe *once more* I corrected myself. As had occurred when I was much younger, I believe I fell asleep. It felt as if I were flying in a dream-like state. So realistic was my vision that I could feel a cool breeze on my face as I rose in the air and looked down at the balconies of other apartments in my block. Although, in this dream, I remained unable to move my body, I could still move my eyes and was able to roll them back to see a blinding orange light that seemed to hang motionless above me in the sky. Although the muscles in my face would not allow me to do so, I felt a smile within me as I rejoiced at the sensation of again experiencing a 'flying dream'.

As I closed my eyes to shield them from the brightness of the light, it seemed as if the light just 'winked out' as I did so. Opening them again, I sensed the feeling returning to my body and the numbness being replaced by pins-and-needles. In my 'dream' I was still not back on the ground though. I appeared to be floating about three feet off the ground. The sensation of not having control over one's body in this buoyant state was quite unnerving. Reaching out to somehow gain purchase on something and steady my movement I wobbled in a very unsteady motion. Being too far away from any area to make contact I reluctantly tried to relax my posture. In doing so I appeared to drop slowly from my floating position and as I did I became aware of a small child sitting cross-legged on the floor a couple of feet in front of me. No more than four or five years of age maybe, she excitedly clapped her hands in front of her face and jabbered a string of words in a very animated fashion: "Fly high in the sky, fly high my Peter Pan!" This wasn't a dream.

My feet settled lightly on the floor and thankfully I regained my bearings. Taking in the surroundings, I saw that the relatively small room strangely resembled the inside of an Eskimo's igloo – the ceiling was domed, whilst the curving walls were bright white and appeared to glisten like ice. Although the room was apparently lit, I was unable to see how – it was just bright! Quickly looking round the perimeter of the room, I was astonished that there didn't seem to be any visible entrance way. How on earth had I entered? The young child continued to babble her rhyme and elatedly clapped her hands in time with the rhythm of her words: "Fly-high-in-the-sky-fly-high-my-Peter-Pan!" She appeared to be dressed in a white nightdress with some sort of animal print covering it. Her feet were bare and protruded from underneath her attire as she sat rocking back and forth in her crossed-legged position. She had long shiny black hair that was pulled back and somehow tied at the back of her head forming a pony tail that bounced from side to side as she moved.

"Who are you?" I enquired softly not wishing to startle the child.

Seeming to ignore my question, she responded with a question of her own. "Are you a boy or a girl?"

Frowning, I couldn't understand why she had asked such an absurd thing. As I knelt down in front of her, my own long dark hair fell down around my face and I suddenly realised the reason for her question. Parted in the middle of my head and reaching down way beyond my shoulders, the length of my hair had obviously confused the little girl. Additionally, whilst there was a need to very occasionally shave a few facial hairs as they appeared, my skin remained quite smooth and pale, only adding to my apparent androgynous appearance. Accentuating the depth and timbre of my voice, I replied with a grin: "I'm a boy, you silly girl". Waiting for a response from the child, I was struck by the brilliance of her eyes; they seemed to 'sparkle' as the light emanated from them like two shiny little stars. As she continued to sing and clap her hands, I was reminded of the children's nursery rhyme *Twinkle, Twinkle, Little Star*.

Lying scattered by her side was a selection of coloured blocks; small cubed shapes made from what appeared to be a translucent glass-like material. As she began to stack the blocks, one on top of the other, I reached out to help her. Brushing aside my hand, the child admonished me, "Don't do that. You can't do it properly". I pulled back and laughingly informed her that she was too bossy for her own good. "It's my trick, not yours," she added with contempt. I watched as she delicately placed three bricks, one on top of the other – yellow, blue and red – whilst sticking out her tongue in focused concentration. She then repeated the sequence and looked admiringly as a stack of six bricks wobbled precariously in front of her. With great intricacy and care, the little girl, with the index finger of her right hand pointing out from her clenched fist, gently pushed the second brick from the top – the blue one. Without unsettling the rest of the stack she kept on pushing until it tumbled from the others and rested in her small left hand. Amazingly the yellow brick above it had remained motionless, somehow suspended in mid-air. Repeating the action, she very delicately pushed out the second yellow block and caught it as it slid out of the stack. My mouth grew ever more open as I sat looking agape at the actions of the tiny magician. There on the floor in front of this animated young child were four bricks – two yellow and two red. The unbelievable thing was this: that the first yellow brick still had no visible sign of support; apparently floating without any respect for the law of gravity. The same seemed to apply to the first red brick that sat floating upon the second yellow brick as it hovered in space over the final red brick, as it remained firmly attached to the floor. Looking up from this spectacle, I watched as the little girl once again clapped her hands with excitement and squealed in sheer delight. "Magic, magic! I'm a magic girl, Peter," she said proudly.

"You certainly are Little Star. You're my magic Little Star."

I didn't make any further attempt to touch the bricks and remained content to watch as she continued to stack and extract the bricks in varying colour combinations. Each time she did so the disregard for the laws of physics was apparent. As a brick was pushed out with her delicate little fingers, the remaining bricks above floated unsupported until she pushed the stack down to join the ones still standing below.

As far as my belief system was concerned what I was witnessing was impossible, and with apparently well-practised ease, I stored the incident away in the unbelievable and unexplainable file tucked safely away in the recesses of my mind.

Finally, she appeared to tire of this particular game and in doing so seemingly lost control of the floating bricks, for it was then that the entire stack tumbled noisily to the floor. As they cascaded around our feet she once again applauded excitedly, filling the air with the exuberant trill of her childish laughter.

Then, before I realised what she was doing, she appeared to reach behind her and produce something new. It looked like a musical recorder, the wind instrument played remarkably badly by a host of school children in ear-splitting recitals throughout the world. It wasn't placed to her lips though, as a normal recorder would be, instead she was pointing it at the bricks scattered about the floor. As she focused in on a particular colour of brick, a humming note was emitted from her 'recorder'. Moving from the red, to blue and then yellow, as she pointed at each brick it seemed to prompt a different musical note. Joining in the performance she pursed her tiny lips and matched the sound note for note. By jumping from one colour to another in quicker succession, an extremely limited, but recognisable tune, began to resonate from the 'recorder' and its conductor. Tiring with this new game quite quickly, she placed her musical baton to the floor, where it continued to produce a string of abstract sounds until it finally quietened.

It was then that I felt an additional presence in the room. Looking up I saw The Witch standing no more than a couple of feet behind me. Concerned that her arrival might alarm the child I had now named Little Star, I turned back to reassure my tiny companion. Although clearly aware of The Witch, she paid her little interest and went back to constructing increasingly high stacks of colourful bricks.

"That's your lady," Little Star whispered as she picked up a red brick and toyed with it in the palm of her tiny doll-like hands.

Surprised by her statement, I asked what she meant.

Without raising her head up, she swivelled her eyes towards The Witch.

"That's not my lady," she said again. "My friend is a big boy."

I looked towards The Witch and wondered what she meant. I had always intuitively known that The Witch was female; purely by the way I felt when I was with her. It was a different matter all together when in the presence of the smaller ETs – the relentless, determined, worker bees of the operation – if anything they appeared to be sexless in their energy, like being in the company of something that, although clearly organic, was overwhelmingly robotic in nature.

"I call her The Witch," I told Little Star, gesturing towards our observer.

"What's a witch then, Peter?"

I went to explain, then thought better of it.

"*What is a witch*, Peter?" Little Star asked again, emphasising nearly each word as if I might be deaf or stupid.

"She's my friend. Have you got a special friend too ... just like her?"

"Yep, but mine is a boy one."

Interesting. Very interesting, I thought.

Whilst remaining really quite mesmerised by Little Star's continued play, I soon became aware that our attention was required elsewhere, although no gesture had been made nor words said to affirm this. Both Little Star and myself stood in unison and turned to face The Witch. Reaching out with her unusually elongated arm she extended her pale, tube-like fingers and took Little Star's small, fragile hand in a flattened encircling palm. Little Star stood at her side and in passing my side reached up and took hold of the fingertips of my right hand.

Looking over and past The Witch, I noticed, for the very first time, some sort of opening in the wall. No more than two feet in diameter, squared-off at the top and probably less that five feet in height, we all three filtered out of this exit. Turning to my left and then my right, what I first considered to be another room was in fact a corridor, curving away sharply in both directions. Although not clear and precise in recall, there was certainly a familiarity to the surroundings that made me feel at ease. Because of the poorly lit environment I was not able to see far beyond where we were standing, but as The Witch took one step to her right, she exposed a seat of some sort jutting out from the wall just behind her a relatively short distance off the floor. It came out about 12 inches from the wall and was roughly 18 inches in width, and appeared to be made from the same substance as the wall, if not integral to its construction. By this point in my examination one thing was undeniable: I knew this location; I had stood here on numerous occasions in the past; and I had sat on that low-level protruding bench.

Without really waiting for instruction or considering my next action, I stepped towards the bench and sat Little Star down beside me. Due to the limited height of the bench, my feet sat comfortably flat upon the floor with my knees pushed up. Little Star very happily swung her legs backwards and forwards with her feet missing the surface of the floor by many inches.

Looking down and for the very first time, I took note of what I was wearing. I had on a short-sleeved, baggy, white t-shirt that had a large picture of a cockerel sitting atop a football emblazoned on the left breast – it was the emblem of the North London football team Tottenham Hotspurs. It hung baggy and low over my hips, slightly covering a pair of black boxer shorts.

Little Star rocked to and fro beside me, occasionally swinging her legs into mine and proclaiming 'whoops-a-daisy' as she did so. Her giggle continued to ripple beside me, although there was no distinct echo as she did so. As her legs swung, I gently leaned into her tiny body perched beside me and pushed lightly, whilst returning the legend of 'whoops-a-daisy'. Our childish play continued on for longer than I can remember, but much too soon The Witch re-appeared from the shadow of the corridor. All the time Little Star and I had sat there playing, there was no real sense on my part of where we were: I had no conscious awareness that I was in the middle of a 'contact experience'. Of course I knew it on some level, but part of me had somehow temporarily suspended my belief system and, as far as I was concerned, I was just sitting somewhere playing with my new little friend. In fact, if I had to put a title to the episode of which I have just described to you, it would be to say that I had been 'babysitting'.

Without warning, Little Star hopped down from her seat and playfully skipped over to the extended hand of The Witch. I remained seated. I had no intention of following her. I knew what I had to do. I had to stay where I was.

Sitting back down on the bench I watched as they walked away. And within a minute or two, I felt my eyelids becoming heavy. A strong sense of drowsiness came over me and before I knew it I was dreaming.

"I long, as does every human being, to be at home
wherever I find myself."

Maya Angelou

Food For Thought: Here, There and Everywhere

Question: Where do the ETs come from and why do they say the things they do?

Answer: On the whole it is believed that ETs are just that: Extraterrestrial. But if they are not, what are the options? Also, there are many theories with regard to why they are here and how they arrived, so I would like to take this opportunity to explore these fascinating questions.

The ETs have never told me consciously anything to do with the above topics, because, surprising as it might seem, I don't remember having ever asked them, and in my experience they don't normally offer up information but only respond to what they deem the 'right questions'. One important thing I have gleaned from my contact, is that wisdom and understanding do not come from being handed information without first discovering the path to that destination. In a way, one has to know the answer to the question already – without realizing it, that is – to enable the appropriate question to formulate in one's mind. I sense that if one is not ready for the truth, one is unable to formulate the question.

As far as I know, in our direct contact with ETs, we are, the Experiencers, more often than not, informed that the alien point of origin is 'far, far, away!' For instance, the famous Contactee* George Adamski (probably the best known alien contact) claimed that the ETs he was in contact with came from several planets in our solar system.

Then there was Truman Bethuram, who one year after Adamski published his book *Flying Saucers Have Landed* in 1953, wrote his own account of alien contact. The planet where his ETs supposedly came from was called 'Clarion' and was apparently 'hiding behind the moon', which is possibly the closest apparent home of the ETs. Also in 1954, another Contactee came forward called Daniel Fry, who was followed by Howard Menger, Buck Nelson, George Van Tessel, Marian Keech, Orfe Angelucci and Antonia Villas-Boas – all claiming that their ET contacts came from either planets within our own solar system or even much further afield. Moving further forward, we have the likes of Betty and Barney Hill, who claimed that in September 1961 they were abducted* and taken onboard a spacecraft whose origin was Zeta Reticuli, a binary star system located about 39 light years away from Earth, situated in the constellation Reticulum. Another candidate for inclusion in this section is Billy Meier, the Swiss farmer, whose ET contacts came from a planet called Erra, in the Plejares star system, which is beyond the Pleiades. Interestingly, the Plejarens stated that they live in a dimension that is a fraction of a second in the future from our own (an alternate timeline that is).

The message of the ETs, as shared with the Contactees*, Abductees* and Experiencers*, is always similar in substance, and I find the meaning behind the actual content increasingly more and more interesting.

The implication of the repeated communication from the ETs is sadly one of doom and gloom with regard to the future of the human race. It is a warning that if we do not stop polluting our planet from industrial sources, raping and pillaging its structure, or threatening its continuation with the detonation of nuclear weapons, we will ultimately kill ourselves and destroy this beautiful and unique world. We have heard variations of that warning since the 1950s, certain words may have changed, as well as the specific targets, but the core of it remains the same.

The question I have is this: why would an alien race come millions of miles, or even several light years, to warn us of a potential calamity occurring to our home, due to the irresponsibility of our own actions, when it would have no apparent impact upon them?

A response to that might be that the ETs are extremely loving and caring (humane – for want of a better word!) and based on the mistakes that they might have made themselves in the past they travel throughout the galaxy looking for other intelligent species who might be in need of advice. Possible? Yes, but is that just wishful thinking on our part?

Another rejoinder might be based on the thoughts behind Quantum Physics and the concept that we all live within a 'field of energy': that is the 'space' between everything and the area that we exist in. The idea is that everything is connected, so the impact of a planet destroying itself would ultimately affect other planets via the 'field'. Could that be possible?! Yes, I believe it could be, and further that there might be a universal 'police force' that oversees us all with regards to avoiding self-destruction and how it might impact the rest of universe.

And then there is also Occam's Razor: that the simplest explanation or strategy tends to be the best one.

Let me run something past you for consideration. It's just a story, a pretend story, but one that might act as a signpost to guide us on to our next destination:

Mr. Bile was sitting at home one evening having a meal with his wife, when he heard a knock on his front door. Opening the door he was greeted by an oriental looking man who smiled and offered his hand to Mr. Bile, who reluctantly took it in his and half-smiled back. This was not the normal behaviour for Mr. Bile, who was generally unwilling and wary to greet strangers at the door, especially ones who appeared 'different' in appearance to him.

The oriental man introduced himself as Dave. He went on to say that he was from "far, far, away", but had "very grave concerns about certain aspects of the neighborhood."

"What might they be?" enquired Mr. Bile, showing a distinct lack of interest in what Dave had to say. After all, he did appear to be a foreigner.

"I noticed that the river that runs through your forest is polluted with such items as bottles, food wrappers and even a metal devise on wheels, which is designed, I believe, for carrying items around your local supermarket before purchase."

"A supermarket trolley?"

"Very possibly."

"What's that to do with me?" asked Mr. Bile, clearly now becoming irritated.

"I don't understand", replied Dave, looking quizzical. "Is it not your home that is being infested with these items?"

"No."

Dave shook his head gently from side to side and furrowed his brow. "Additionally, I must bring to your attention that many acres from the said forest have been savaged to make way for a range of commercial structures."

"Do you mean the new shops? You can't have too many McDonalds or Starbucks can you?"

"But much wildlife has been destroyed my friend, as well as natural beauty."

"Why on Earth am I continuing to listen to this fool?" thought Mr. Bile.

Dave carried on. "I just witnessed a neighbor of yours haphazardly throwing several bags, apparently containing household waste, up and down the road."

"Don't worry about that; the refuse collectors will probably deal with it or somebody else will clear it up."

"I also watched as your children threw waste from food products onto the gardens." Dave waited for some sort of response from Mr. Bile, but none came. "And your cars continue to emit gaseous waste from their exhaust systems, which pollutes the air that you breathe. Does this not concern you?"

Mr. Bile's eyes widened. "Are you collecting for something or are you a Jehovah's Witness? You're a bit foreign aren't you – maybe even an illegal alien?"

Dave was not smiling anymore. "No, nothing about me is illegal or alien my friend."

Closing the door, Mr. Bile said finally, "Not today thank you."

Mr. Bile told his wife all about what had happened.

"If he is from far, far, away?" Mrs. Bile said indignantly. "Why is he so concerned about what we do around here? What's it got to do with him anyway?"

"I don't know. Just one of those interfering do-gooders, I suppose," replied Mr. Bile, going back to his artificially flavored, pre-packaged, processed and genetically modified meal.

A couple of days later, Mr. Bile came back from the supermarket. He said that he had seen that 'foreign guy' Dave. Disapprovingly, he told his wife that he had watched him at the customer service desk, apparently holding everybody else up from buying their cigarettes and lottery tickets.

"What was he doing?" Mrs. Bile asked.

"He wanted to know why he couldn't buy locally produced food in the store, because all the vegetables came from overseas, and why all the meat contained so many additives, and the chickens were from battery farms. Also, he was complaining about the amount of packaging with every item and wanted to know what they were going to do about it because there was excessive waste everywhere."

"I told you didn't I, just a complete nutter," Mrs. Bile said, rising from her chair and clenching her fists in anger. "He has absolutely no right to interfere with how we live our lives around here. He should go back to where he came from... *far, far, away...* and make sure he stays there!"

"Well that's the thing, somebody else in the queue said he was local, that he had lived around here for years and years!"

Considering the above, if the ETs are from 'far, far, away', as they claim to be, why are they so concerned about what we are doing to this planet and ourselves? Suspending any possible belief in the two theories I put forward earlier – their possible "love and care for the human race" and the concept of a "quantum field of energy that connects everything in the universe" – would the Occam's Razor evaluation of the situation not suggest that they might actually live a lot closer to where we live? Would it not be worth consideration that Earth might actually be their home in one shape or form? And taking that idea one step further, what about the possibility that Earth might actually have always been their home? After all, it might be worthwhile to raise the question of who are the landlords and who might be the tenants?

It is also true to say that there are many, many theories regarding why the ETs are here. First we will approach the question from the angle that they are visitors to this planet and do actually come from 'far, far, away'. What would possibly be the reason for travelling millions of miles to visit us? Are we that interesting? In my opinion, I would have to answer in the affirmative. Maybe the human race is a unique species and we are worth observing. Alternatively, could it be that the planet itself is the thing of interest and the reason in coming here has nothing to do with us? Maybe the actual structure of the planet – the minerals contained within – are worth the journey? But that theory would need to rule out the fact that we are here already and the ETs would have to have no concern about our current and already lengthy occupancy. Does that sound like anybody else you know? Does it make you think of certain human world powers who have previously entered – uninvited – independent countries, only to begin dictating how they are then managed whilst pilfering their natural resources?

There is also the possibility that the ETs could just be tourists, out on an extended cruise of the galaxy taking in specific sights along the way. There are many possibilities and any of the ones I've mentioned could be true and might have happened at some point in the past.

Another theory worth considering is that the ETs might actually be 'time travellers' – human beings from Earth or another race of beings from the future. I guess that all of the above theories might even fit in with this one. Just change 'time traveller' for 'traveller from another dimension' and you have yet another theory to add to the chain.

Now, I truly believe that throughout the history of this planet we have definitely been visited by a race of alien beings that originally lived on a planet 'far, far, away'. I also believe that we have been visited by beings from other dimensions and from a different time.

However, by applying Occam's Razor once more, it is my belief that a high percentage of the 'flying saucers' we see in our skies might just come from a place much closer to home: a home that is called, by the beings that fly in these fantastic craft, the planet Earth. As for the lineage of the beings, I would have to say that we are, more than likely, closely related to them. In fact, you might say we are cousins.

I believe that several different types of humanoid beings seeded the planet Earth millions of years ago. It is no coincidence that we all look so different from each other – it really cannot be claimed that the climate changed our appearances so drastically from continent to continent, although it might be true for a small percentage. Therefore, based on that premise, *we are all ETs.* Some of us could probably be Martian or maybe Venusians. Maybe the origin of others is even more exotic, possibly from another star system, even the Pleiades perhaps.

I suggest that the history of the planet Earth reaches further into antiquity than we have even begun to imagine. Only a relatively small amount of that history has actually been recorded and saved and then often only if it 'fits' our current belief systems regarding the evolutionary process. What does remain, apart from the records pertaining to the last few thousand years, is to be found in the mysteries, legends, fables and folklore of ancient societies and civilizations around the planet. Invariably these legends say the same thing: *that our ancestors came from the stars.*

The human race is very old – *older than any of us can even imagine* – and our origins might not necessarily be from this planet, although I am sure certain facets are. Therefore, it is vital for us to consider these possibilities when asking the questions raised here.

One year I traveled back from the UFO Congress conference in the minibus that ran from Laughlin to Las Vegas. My fellow passengers included Budd Hopkins and David Jacobs, both Abduction researchers and published authors on the subject. The question was raised in conversation with regards to the whereabouts of the ETs when they were not traversing our skies in their impressive modes of transport. "Where do they go?" asked one of the other passengers. David Jacobs responded by saying, "They're hiding behind the scenery". Based on my particular belief system with regard to the domicile location of the so-called ETs, I considered it to be rather an astute observation, although I'm not sure if Mr. Jacobs realised.

With regard to the science of travelling to this planet from 'far, far, away', it certainly does appear to be beyond our current understanding of how the universe really works and is limited by our known knowledge of physics. Interestingly though the following statement made by Dr. Harold Puthoff (Director, Institute for Advanced Studies at Austin and author of *Fundamentals of Quantum Electronics*) makes interesting reading: *"The possibility of reduced-time interstellar travel either by advanced extraterrestrial civilizations at present or ourselves in the future, is not fundamentally constrained by physical principles."*

So, just because our scientists say that it would be impossible to travel to Earth from another star system, their assumption based on our present day understanding of propulsion and travel, does not mean that this state of affairs will always be so. There are many examples throughout history to illustrate how we are limited by the time period in which we find ourselves and by our understanding and acceptance of what could be possible in the future.

If the ETs are travelling from the future or from a different dimension, it is safe to say that the marvelous craft we see speeding through our skies come from an advanced science that is presently beyond our comprehension. And clearly the ability to get from 'there' to 'here' also comes under that umbrella of appreciating the possibility of the 'impossible'. After all, a modern-day 'soothsayer' once said: "If an elderly but distinguished scientist says that something is possible, he is almost certainly right; but if he says that it is impossible, he is very probably wrong." That established, proven and highly respected visionary was the science-fiction writer, inventor, scientist and 'futurist' Arthur C. Clarke.

There is also a great deal of speculation about what occurs during the so-called abductions with regards to medical examinations and procedures, such as the extraction of sperm and ova, and other occurrences that many believe are linked to the idea of producing some sort of alien/human hybrid. There have also been times when it has been suggested by Investigators that alien/human-looking babies I encountered were 'connected' to me in some way. So, based on what I have personally witnessed, I believe that there is some sort of biological programme in action, although there is something inside of me that remains positive about these events, even if I can't quite put my finger on why. I believe that our contact with these Beings is a positive one. I feel it in my bones. So I am unable to support the negative theories proposed by some investigators concerning the reason behind this apparent cross-pollenisation of '*them and us*'.

Some people believe that the ETs do not have any feelings, and based on the remarks I have made regarding some of my experiences, one might be led to agree. But I don't believe that is true. My take on it is this: the ETs have a very good understanding, appreciation and knowledge of emotions and feelings, but the difference between us and them is that they are not controlled by their feelings.

On the whole, I believe it would be fair to say, that the majority of our problems here on Earth are based on our inability to control our feelings. The ETs might appear to be emotionless and cold to somebody outside looking in, but that is not true as far as I am concerned: I have experienced a deeply profound love when in the presence of some of them. It's true that the smaller Greys project a disconnected emotional attitude when dealing with humans during an abduction experience, but they have never hurt me in any way. I might have been scared, or believed I was scared, based on my corrupted belief system at the time, but I have never been physically harmed. There have been times during certain procedures where I have felt uncomfortable, just like one would during a normal medical examination where it hasn't been made clear to the patient what is going to happen. After all, the unknown can sometimes conjure up a feeling of instability or exaggerate a slight sense of insecurity. There have been no broken bones; no long lasting trauma that needed repair or anything like that. In fact, virtually nothing has ever occurred that I haven't previously experienced in a professional human medical environment.

"The laws of physics are merely suggestions."

Unknown author

Experience: A Night at the Disco

London – May 1976

The memories from this year remain very vivid: it was a time of extremes, in one shape or another. I travelled between the polar ends of my complete emotional range, yet despite it being such a time of highs and lows, I must say that what transpired played a major part in what would eventually lead me to some extreme life changes by the year's close.

David Bowie returned to these shores after having been a 'soul man' in America the year before and arrived, transformed once more, at Victoria Station in London giving a (apparently) Nazi salute to the waiting throng of adoring young fans. Sadly, just before he arrived, he had given an interview saying that the UK could benefit from a fascist government so if the 'salute' had really happened or not was neither here nor there as far as the British tabloids were concerned. "The return of the Thin White Duke, throwing darts in lover's eyes," sang the boy from Brixton as he toured the British Isles that year.

As for dear old Blighty, we remained under the constant threat of attacks from the Provisional IRA with bombs still being detonated in central London. This situation was not helped when the Northern Ireland Constitutional Convention was formally dissolved resulting in direct rule of Northern Ireland from London via the UK Parliament. It makes you wonder if the people who make life-changing decisions sit down and say: 'What could we possibly do to make this situation worse?'

Throughout my teenage years, more often than not, I had a regular girlfriend. Although I still maintained a friendship with my friends from school this would only take up one night a week and the rest of the time would be spent in the embrace of the fairer sex. I just preferred it that way; I had never been what you might call 'a man's man' and I remain that way to this day.

On one occasion, when I was in-between romances, whilst attending the regular Tuesday night gathering at the pub with my drinking chums I mentioned that I was available at the weekend if any of them had any plans. Out of the group there was a small circle of about four who never seemed to have a girlfriend so it was with these particular lads that I hooked up with on that weekend and a few weekends that followed.

During my brief, enforced hiatus from all things amorous, my friends and I would normally decide what we were to do at the weekend on the Saturday. Sadly, their idea of a good night out would be an evening in a discothèque spent guzzling over-priced, flavourless lager, whilst straining to hear each other speak over the pounding racket of emotionless disco music. Although their intention never appeared to actually meet up with any girls at these locations, they were nonetheless extremely eager to attend such gatherings where there would be a surplus of single females. I could never work that out. I guess they were just shy young lads with very little experience of the fairer sex. Of course, I was very different when it came to approaching possible candidates to rescue me from this ongoing male-dominated social lifestyle. On more than one occasion though, one of my friends would put the kibosh on any potential romantic interlude I was setting up. I could never work that out either. Maybe it was their lack of experience with the ladies that compounded their inability to approach them and ultimately fulfill any romantic plan they might like to have. Also, did they feel that if they couldn't have a romantic liaison, they would make sure that I couldn't either?

One particular Saturday we had arranged to meet up at our normal drinking hole and proceed from there. Several ideas were bandied about until we finally settled on a new venture: the Leigham Court Hotel, a local hotel that had a relatively small disco attached to it, and although not much was known of this location it was our best option as we didn't really feel like going into central London and spending too much money. As we pulled into the hotel car park that evening, absolutely nothing we saw encouraged us to believe that we had made the right decision. The hotel itself probably struggled to manage two stars in any guide book, whilst the sad little lean-to structure that was precariously attached to the side of the main building and bore the uninviting garish red, pulsating neon light above its entrance, boldly stating 'Welcome to the Disco' left a lot to be desired. Upon entering, and surprisingly, there was a sense that this was very much a private gathering. All heads turned as we hesitantly sauntered across the tiny dance floor on our way to the bar; looks of apparent disappointment from our audience as we were clearly not recognised. The sign, which hung over the entrance, did not live up to its statement; the atmosphere was really peculiar and left you feeling most unwelcome without even a word being said.

The room was relatively compact for a venue of this nature and you felt more like being in a village hall during a family wedding. It did not contain the regular fixtures normally found in a disco of the mid 1970s: no spinning mirror ball, pulsing disco lights, comfy intimate seating and charismatic, cocktail-shaking bar staff. It all seemed more than a little ram shackle. I felt disappointed and very ill at ease.

Securing a group of seats in the corner of the room, we sat and sipped at our pints of beer. The music, loud and unrecognisable, boomed from an inadequate sound system that distorted with every crackling beat that vibrated from its speakers. Continuing to make awkward and uneasy eye contact with those who sat around us, I felt that it probably wouldn't be long before we left. As usual, the decibel level did not allow normal conversation so I could only imagine if my friends were feeling the same way, but by the looks on their faces I believed I wasn't alone in my experience.

And then it happened. Imagine watching a movie and all of a sudden the movement of the characters went into slow motion; still moving but virtually imperceptible. Of course, I noticed the lack of sound immediately: from a deafening vibrational thud to a stunning vacuum of silence. I thought that everybody else had been caught out by the moment until it finally dawned on me that each and every person appeared to be frozen in time. I stared in astonishment, watching as glasses were very gradually moved to waiting lips and tilted to dispense the liquid within. Dancing feet, which had been caught in mid movement, were placed, excruciatingly slowly, back to the dance floor. Gesticulating arms returned to stationary positions and closed blinking eyes reopened. This strange scene continued to play out as I rose, without even thinking, from my seated position. I was moving normally or so it appeared to me. Moving naturally, at the right speed.

Looking back at my friends, apart from the fact they too were moving in slow motion, they seemed quite normal: caught in conversation, drinking, looking around them. I had a profound feeling that I had to remove myself from the room, especially as I was finding it hard to breathe, as if there wasn't enough air in the room. I began to move, each step laboured as I experienced a definite resistance as if I was wading through water. Locating the exit, I moved with some difficulty towards it. To my right-hand side was the bar, where a number of people stood in the process of purchasing drinks. A tall black man with obviously dyed short white-blond hair had already turned towards me and was about to start drinking from a pint of Guinness that hovered by his lips. As I walked by him the glass had imperceptivity moved away by an inch or two, leaving a frothy line of foam upon his upper lip that strangely matched the colour of his hair. There was a very brief moment where I thought he acknowledged me; just the stare of somebody in a painting when their eyes realistically connect with yours, but after another step I realised that I must have just stood in his line of vision for a fraction of a second. On my left and just by the exit door were two girls, one bending down to place her handbag on the floor, whilst the other girl had her hands raised above her head in response to an unheard rhythmic beat. I reached out and pushed the exit door. In another step I was into the foyer, walking past a ticket booth, a small group of revellers on their way into the club and two dinner-jacketed bouncers. Navigating with great care through this little huddle, it was now apparent to me that I was probably not visible to anybody at all. Finally I stood before the frosted glass doors which bore the name of the club in reversed lettering. I hesitated momentarily, as something did not appear quite right. The light piercing through the doors was too intense for this time of night; it felt like daylight. I grasped the bronze handle and pushed. Stepping out onto the pavement beyond I automatically shielded my eyes from the unnaturally bright light that greeted me. Any hope of a rational response had left me some time before; I was an observer who sought no explanation. A group of young men and women stood at the kerb by a line of parked cars, dramatically illuminated by the abnormal light source that made everything appear with pronounced, sharp definition. Raising my line of vision without removing the shield of my hand I could see that the light was not coming from the surrounding lamp posts; in fact the source was much higher and was focused in one particular area. Increasing the angle of my hand for better vision, I went to look towards the light and then everything went black. As if somebody had just flicked off the light switch, I stood in complete darkness, not being able to see anything at all. And then once more it was light, but I was now no longer where I had been.

You have to remember that what then transpired happened in the blink of an eye. My head was still tilted at an angle as if looking up although I was no longer standing, but lying down. It took a moment or two to take on board where I was, but my brain soon clicked into recognition: I was lying down in my bedroom at home staring up towards the light that hung from the ceiling. It's totally amazing how I could so easily embrace and hold on to a reliable and secure state of mind, whereas a moment before I was struggling to remain mentally stable in a surreal dream-like condition.

I just lay there staring. I was being bombarded with the familiar and the safe. I guess I was in fight or flight mode. The memory rushed away from me like a departing train, gaining speed with every second. Somewhere in my brain, as part of the fight or flight programming, a chemical was released or an electrical impulse was sent out, because within moments my eyelids were drooping and an overwhelming sense of drowsiness was taking over. Before I could think or evaluate anything further, I fell asleep. Just like that, I fell fast asleep.

I don't remember dreaming, I just woke up. I felt rested. In fact, I felt great. It was morning. Looking at my alarm clock I saw it was late morning. Within a few moments of waking there was a knock at my bedroom door and I heard my mother announcing that I had a visitor downstairs. I scrambled to get dressed because I knew that my visitor, whoever it was, would have been left on the doorstep and not invited in! Opening the front door, my friend, Robert, was waiting. We'd been out the night before hadn't we? The first thing I noticed was that he had a rather angry looking black-and-blue eye. Naturally, he didn't appear to look too happy.

"What happened to you last night?" he asked.

Yes, we had been out together. Where had we been? My mind struggled to recall. How much had I drunk last night to have this sort of memory loss? While I was still trying to remember, I enquired what had happened to his eye.

Robert winced slightly.

"I got a major slap as we were thrown out of the club."

No, I still had no idea what he was talking about or how I might have been involved.

"Where did you go last night? We were all really worried. We thought you must have been beaten up someplace." His face was still clearly troubled from the memory of it all.

A brief snapshot of being somewhere last night flashed up before my eyes, but nothing more. My brow furrowed and I shook my head.

"Blimey mate, I've got no idea. In fact, I can't even remember getting home last night. I can sort of remember being somewhere and... it wasn't much cop...but that's all", I said struggling to remember more details.

Robert narrowed his eyes. Maybe he didn't believe me I thought, but my sincerity must have convinced him finally because the look on his face changed.

"You sort of just disappeared at one point, pretty soon after we arrived", Robert continued finally. "One minute you were there and then the next – bosh! – you were gone."

I nodded, but I didn't really know what he was talking about.

"Anyway, soon after that everything just kicked off." Robert narrowed his eyes again; maybe it was the pain from his eye. I just shrugged and he went on.

"We were just having a drink and then the bouncers pounced on us, giving us one almighty slap. They dragged us outside and laid into us. Some people came out with them and were shouting that we weren't welcome and shouldn't come back. Couldn't believe it?"

I shrugged again. I now had a vague memory of arriving at the club last night, of feeling a bit uncomfortable and then... nothing. I had woken up this morning and that was that. I said all of this to Robert and I asked him if he wanted to come in, but he said that he had stayed at Richard's last night and was now going home to his place. I told him to take care and that I would speak to him later. He winced one last time, turned and walked away.

Clearly, I now have better recall of what happened to me on that night, but only as far as I have shared with you. Something strange went on, that's for sure. I have no facts to support this theory, but in some way or other I sense that I was being *protected*. By whom I'm not sure (but I have a good idea), but whoever it was clearly knew what was going to transpire and manipulated my surroundings to retrieve me from any potential harm.

"Human intelligence is richer and more dynamic than we have been led to believe by formal academic education."

Sir Ken Robinson

Food For Thought: What is Intelligence?

For some time now I have been fascinated with the subject of intelligence. Not just the intelligence of human beings, but of the potential intelligence of all living beings upon this planet, including areas not normally associated with the concept of intelligence, such as plant life, and even the planet itself. My interest has now grown beyond the confines of this world, obviously due to my contact experiences with a so-called 'alien' intelligence that would appear to exceed that of our own.

I would imagine that a lot of people would respond in a similar way when asked what they believe human intelligence is. However, the answers might be somewhat mixed when asked if animals, plants or the planet might also be considered intelligent, in comparison to human beings that is.

To begin with we need to understand what intelligence actually is. Where do we start? I suppose our starting point is always based on the human viewpoint: the anthropomorphic* belief. An animal can't operate a computer; therefore it cannot be intelligent, is probably a sound example. A plant is unable to talk; therefore it is not classed as intelligent. The planet is just a big round ball of stuff; clearly it would be ridiculous to say that it was intelligent.

So as to keep a lot of my research close to home, which would enable me to thrash out and understand the views easier; I discussed the question of what intelligence is with anybody in my immediate circle who would listen. Here are just a few of those responses, varying from a key-hole view to a much more expanded perspective:

"It's the ability to comprehend, to understand and profit from experience."

"Ultimately it's something that's exclusive to humans: it makes us smarter than everything else."

"Intelligence is the capacity to reason, to plan, to solve problems, to think abstractly, to comprehend ideas, to use language, and to learn."

"It's something I wish I had more of."

"There are several ways to define intelligence. In some cases, intelligence may include traits such as creativity, personality, character, knowledge, or wisdom. However there is no agreement on which traits define the phenomenon of intelligence."

"Intelligence is what makes me *me*."

To believe that intelligence is limited to the human race is nothing short of humorous. When it is finely discovered that the human form of intelligence is but a seam within a rich vein of universal intelligence, the concept of believing otherwise will be seen as wonderfully absurd. But for now that form of hilarious absurdity can only be appreciated by a relatively small amount of intelligent beings, and I know one particular Aspidistra that regularly laughs long and hard.

It would be true to say that intelligence is a term that is difficult to define, and it can mean many different things to just as many people. In fact, it has divided the scientific community for decades and controversies still rage over its exact definition and form of measurement, which is encouraging I have to say. In the popular sense, intelligence is often defined as the general mental ability to learn and apply knowledge to manipulate your environment (I wonder does that also include polluting and ultimately destroying one's environment or even blowing it up with a nuclear weapon?). Other definitions of intelligence include the ability to reason and have abstract thought, adaptability to a new environment or to changes in the current environment, the ability to evaluate and judge, the ability to comprehend complex ideas, the capacity for original and productive thought, the ability to learn quickly and learn from experience and even the ability to comprehend relationships. But one must never forget, although these definitions are eloquent and articulate, they have always come from a human viewpoint based on the concept that these are skills displayed by intelligent human beings. In short, I would have to question the starting point for coming up with these descriptions. A superior ability to interact with the environment and overcome its challenges is often seen as a sign of intelligence. In this case, the environment does not just refer to the physical landscape (e.g. mountains, forests) or the surroundings (e.g. school, home, workplace), but also to a person's social contacts, such as colleagues, friends and family – or even complete strangers. Researchers, asked about the aspects of intelligence, felt that factors like problem-solving ability, mental speed, general knowledge, creativity, abstract thinking and memory all played important roles in the measure and standard of intelligence. Most agree that intelligence is an umbrella term which covers a variety of related mental abilities. Psychologist, Robert Sternberg, proposes that there are three fundamental aspects to intelligence (*Triarchic Theory of Intelligence*, Sternberg, 1985, http://tip.psychology.org/stern.html): analytical, practical and creative. He also believes that traditional intelligence tests only focus on one aspect – analytical – and do not address the necessary balance from the other two aspects. One alternate type of intelligence often mentioned in popular media is 'emotional intelligence', developed by many researchers. This refers to an individual's ability to understand and be aware of your own emotions, as well as those of people around you. This ability enables you to handle social interactions and relationships better. In the educational context, a person's intelligence is often equated with their academic performance but this is also not necessarily correct. Certainly, a person's ability to think analytically and use their knowledge and experience is often more important than their ability to command a large number of facts. Note also that the word intelligence comes from the Latin verb *intellegere,* which means 'to understand'. However, the ability to understand could be considered different to being 'smart' – the ability to adapt and be 'clever' and the ability to adapt creatively! I am intrigued by many of these theories and would like to suggest that several aspects could be so easily attributed to the known behaviour of some animals. Is it our narrow-minded belief system (or plain old fear-based response) that stops us from embracing the possibility that intelligence is manifold and not confined to the exclusive domain of the human animal? Why do we appear scared to consider this possibility? In our place as the only recognised creature with intelligence on this planet, are we so insecure in that position? What is hiding deep within our psyche that encourages such anxious behaviour? Surely this is not the action of an intelligent being?

I know I keep coming back to this point, but I feel it is probably central to our discussion: human thinking is essentially anthropomorphic. We measure intelligence against our own scale. Yes, we have developed devastating technology and we continue – and rather scarily it must be said – to manipulate our environment, and no other creature on Earth has achieved that, but which one of us, based on that, would be deemed to be intelligent? As abstract as it might sound, we might consider setting aside our own human way of thinking and try some alternative approach that could possibly stretch our capabilities and help us escape from our self-enforced intellectual imprisonment. If, for instance, intelligence was based on the act of surviving as a species, how far up that particular ladder of evolution are we? There are countless species that would be placed above; although dinosaurs are now extinct, they did exist for million of years, did they not? Because of the way we use technology like errant schoolboys striking matches with no thought or awareness of the danger and the insensitivity we apply to our environment, we may not survive very much longer in comparison to the giant land creatures that once dominated this planet. Perhaps, based just on that behaviour, it would be wrong to classify ourselves as intelligent? What if the true sign of intelligence was to free ourselves from the current restrictions of our life style; if we were to become less dependent on toil and task for instance. If this were the prerequisite to be deemed intelligent, then once again I ask you where would we fare on the ladder of evolution? If we are looking for a candidate within the animal kingdom that could fit that bill, we would have to look no further than possibly the dolphin. As far as we can tell their life is made up of play and leisure (although there has been rumour of some dolphins having been observed behaving outside of this generally accepted conviction of the species, but that remains speculation at this point), a lifestyle that could be described as one of constant artistic expression through their movement within their environment of water. The above thoughts, although abstract and more than a little fanciful to the ears of some people, should not be dismissed out of hand just because it does not fit with previously defined ideas on intelligence. I am aware that a lot of what is proposed here is speculation on my part, but there may actually be research out there to prove my point, but that is really not the central thrust of discussion at the moment. My point is that we don't really know how *we* use our own brains, let alone understand how other creatures use theirs, which will probably turn out to be radically different to ours. Elsewhere in this book I've produced information concerning current theories on how our brain works, but I will add that it is believed we have two very separate hemispheres to our brain: a left side that represents the logical and sequential process and which deals with reading, writing, numeracy, and analysis; and the right-hand hemisphere, which deals with the artistic, creative, intuitive, holistic brain. With this side we comprehend the world via images, colour, music, and spatial relationships. Apparently our seat of emotion is based on the right-hand side of our brain. Although activating the brain for business purposes uses the left hemisphere, training is largely centered within the opposite hemisphere so as to stimulate growth in more abstract approaches to the corporate sector. If anything it might be fair to say that our scientific learning might have been hampered in the past because of our dual brain hemispheres and our fear of 'pushing the envelope'. Because we continue to focus on the attributes of the left brain to indicate intelligence within our species, has our overall spiritual and intellectual growth been stymied out of our lack of an expansive vision? Maybe the dolphins have mastered the ability to draw from what is really needed to be intelligent? Clearly they do not need science or technology, but the fact that they may have an enviable and more evolutionary sustainable lifestyle based on a different behavior than ours may mean that they have developed their intuitions and creativity. If this is true, how on Earth are we ever going to establish any form of communication with them or even begin to appreciate their concepts with regard to the how and why they live their lives?

Although I have experienced a type of communication with a certain kind of ET, could we ever really understand their way of thinking? If their brains have developed in a radically different way to ours, the potential gulf between our two cultures might be ultimately unbridgeable. I wonder what effort might have been exercised to actually communicate with me and millions like me? This I suggest is the real proof of intelligence, because it has proved impossible for the human race to have even one genuine communication with any other form of life on this planet. Having said that, if we have no productive and informative idea of what intelligence is and how it works then we will never be able to establish reciprocal dialogue with any other species – on the planet Earth or off it. Could the fact that when I am in communication with the ETs, I don't experience language as I do with fellow humans – yes, it is heard, yet not in the normal sense – shine any light on our understanding and appreciation of intelligence? Is that how they communicate all the time? Between themselves and with other alien species? Are they somehow 'dumbing down' in order to communicate with us? Is their form of intelligence reflective of ours when it comes to our inability to communicate with dolphins or any other life form other than human beings? But they are making themselves understood are they not? So are they surpassing that barrier by stimulating within us an ancient form of interaction now redundant in our species? Is that then a definition of intelligence: to be able to communicate with a different species? Continuing to think along those lines, could language actually be a barrier to identifying intelligence? Just because we don't have a working language with animals, does that make them less intelligent than us? Do they even need to have a dialogue with us or is that just our perceived agenda? If a lion or a tiger was able to speak English as we do, would we be able to have a productive exchange, considering that their concepts of life and their belief systems might be so 'alien' we would have no way of understanding each other? We only need to look at the problems of communication we have with each other to appreciate the difficulties, especially when you bear in mind the different mindsets, beliefs, cultural outlooks and so on. Obviously this is not the same as trying to talk with a species that has such an 'alien' mind set, but it illustrates how difficult it can sometimes be to understand and appreciate the thoughts and beliefs of others. I have come to the conclusion – in part anyway – that intelligence is actually many things and that our understanding and appreciation is limited to say the least. Not until we can communicate with each other affectively and productively can we hope to be able to communicate with an 'alien' culture. And, if and when we can, perhaps then we can call ourselves intelligent.

"Silently, one by one, in the infinite meadows of heaven,
blossomed the lovely stars, the forget-me-nots of the angels."

Henry Wadsworth Longfellow

Experience: Down on the Farm

Devon, England – August 1976

In August of 1976, I was nineteen years old and what had been a platonic relationship with the sister of my best friend had now blossomed into something more romantic. Although I wasn't to know it at the time, fate did not intend it to last beyond the year's end, but its fond memory would remain for many reasons. Although my friend wasn't overly keen with the arrangement, for whatever reasons that protective brothers have, my new girlfriend, Sarah, and I were very happy together. Sarah's family had planned to go to Devon on the southwest coast of England for a couple of weeks and they had very generously invited me to join them. We were going to stay in a cottage that was attached to a working farm. The party was made up of Sarah's mum and dad, her sister and her husband and her brother (my friend) and his wife. Cars were packed to the hilt and off we trundled in a southerly direction down the motorway. Sarah was two years younger than I and we had already known each other for a year or so before we started the relationship. We thoroughly enjoyed each other's company and whilst in Devon we spent a lot of the time immersing ourselves in the day-to-day running of the farm: milking the cows, rounding up the sheep and feeding the horses. For two city kids it really was an education and a wonderful experience to share together. Most evenings were spent in the preparation of the evening meal with everybody sharing the workload. After the cooking paraphernalia was washed and returned to its storage location we would sit outside in the garden behind the cottage and take in the beautiful vista of the surrounding farmland, thankful for the warm and clement weather. As the evening grew darker the night sky would become a black velvet blanket speckled with a million tiny pin-pricks of diamond white. Sarah's mother had a serious heart condition and after dinner she had to ingest a colourful handful of prescribed medication. One of the side affects of the tablets was that she would become drowsy and could not stay up much longer after she had gulped them down. On one particular evening as the holiday party sat on a selection of outdoor chairs and blankets serenely taking in the majesty of the night sky, Sarah's mother mumbled something about one of the stars winking at her. Looks were exchanged within the group as if to say that maybe it was time for her to go to bed. Pointing skyward, she gestured to a specific area and said that one of the stars was signalling to her. Following her gaze, I looked towards the star in question. It appeared to twinkle with a slight rosy glow, but apart from that looked no different from its neighbours. When the mother made another comment about its luminosity, the rest of the family also focused in on that point in the sky. If anything, I thought, it had gotten a little bit bigger or maybe it was just somehow brighter. As I gazed at the star it seemed to pulse and in a peculiar sort of way I could feel its rhythm. I raised my hand up until the tip of my upturned thumb obscured it. I stood motionless, feeling the slow rhythmic pulse apparently moving through my hand as if I was somehow connected to this distant light source. "I can make it move," I heard myself say in a slow hypnotic voice. A few people giggled as they clearly thought I was kidding around. With just a little more emphasis, I repeated my claim and as I did so I watched as the small point of light emerged from behind my thumb-tip and moved very slowly to the left. Everybody gasped and watched astounded as the star moved in a horizontal line from its original position. After a few seconds it stopped. People were looking from the star, to me and then back to the star. I felt as if an invisible thread connected me to the star. With my hand still outstretched, but no longer obscuring the star, I spoke once more. "Please move back," I instructed softly.

No sooner as the words had left my lips the stationary star began to slowly move once more, retracing its passage and finally returning to its original position obscured by my thumb. My arm dropped to my side. As the star had come to a final stop, its brightness increased dramatically and burst like an exploding firework. As each radiating spark dissipated it was clear to everybody that the star was no longer there. We all stood in silence looking at where it had been.

As the star had apparently and dramatically switched itself off, I immediately felt the connection broken; my focus and attention returned exclusively to the moment. Nobody said a word. Nobody moved a muscle. As if caught in a moment of time with no way of moving on from it, everybody seemed fixed to the spot, both physically and mentally. What had just happened?

When recounting the matter at a later date, I recalled that the holiday party eventually moved away from the spot and, without talking about what had just occurred, went directly to bed. I didn't remember the incident being discussed the next day. In fact, the rest of the holiday went by without further incident and we eventually returned to London all the better for our sojourn down on the farm. Of course this could have been an ordinary event, although, I am quite sure that it was something that we were linked to and had an influence over in some way. I believed that the star was more than just a normal heavenly body, that is was an apparition of greater significance and under intelligent control. But the greater question by far, for me, is what did it mean and what was its purpose? I now understand these type of events to be so much more than an unexplainable light in the sky. In my heart I know that these occurrences have a profound meaning and are provoking us to respond with questions and not just wonder. This has always been true: ever since the beginning of the modern UFO era in 1947 we have constantly been given the opportunity to reflect upon who we are in all of this. In my view, the proverbial unidentified flying object and their apparent occupants have only ever been the Universe's way of nudging us in the direction where we can begin to look at who we are and what our real purpose is in this journey of life. Are we being told a very simple message; that we are far greater than we believe ourselves to be and that the magic of life is actually a wonder to be found throughout the cosmos? Beyond the obvious of what transpired for us all sitting in the garden that evening, what unseen influences were at work and what impact did the events have on our individual psyches? How were we different after that evening together than we would have been otherwise? In all things we need to consider looking beyond our given horizons, because by limiting our view of anything we might be hampering ourselves from seeing what was really there all along.

Part Three

The Adult Years

"ET contact is the greatest story in human history...
the greatest story <u>of</u> human history."

Retired Commander Sergeant Major Robert Dean

Experience: The Fish Tanks

Location unknown – 1977

I've always been confused, *and* fascinated, by the way that Great Britain 'took to the waves' in the distant past and went about ceremoniously planting their country's flag into any landmass that took their fancy. Considering that the initial settlers to what eventually became the United States of America were primarily British, if not European, it's easy to see this mindset at work. I suppose the cherry on the cake of that particular colonial 'land grab directive' had to be the placing of the Stars & Stripes on the surface of the Moon. What was that all about?

How do you think it would work if we – a supposedly highly advanced civilisation – landed on another planet and decided that we were going to take ownership of it, only to find out that it was already occupied by an indigenous race of beings? Maybe it has already happened; after all, the whole Apollo/Moon Landing subject of 'did they or didn't they go to the moon' raises an awful lot of questions that have never been satisfactorily answered.

So, archaic as it might seem in this so-called modern world, 1977 saw Elizabeth II, the queen of Great Britain, tour the world in the year of her jubilee, visiting outposts of the Empire which most school children would have a problem identifying, let alone spelling or even pronouncing their names.

During the extremely hot summer of that year, we here in London complained *and* rejoiced in the high temperatures that brought us all out into the sunshine and saw most of us turn from a whiter shade of pale into a distressed burnt lobster red. With the good weather came a contact experience for me that would write a whole new chapter in my ongoing story and would begin a fascinating train of thought raising innumerable questions that remain unanswered to this day. Not wishing to take up space reiterating the lead up to the main focus of the experience, I will press the fast-forward button and take us beyond the familiar details of initial contact and the transition that occurs from *my* location to *theirs*.

Finding myself in the recognisable surroundings of the 'igloo room' – the white domed environment where many of my contact experiences have occurred – I stood conscious, relaxed and aware of my circumstances in the presence of my regular ET host, the being whom I refer to as The Witch. She stood before me, her right arm rose up and she almost imperceptibly brushed my left elbow with her hand. In doing so I realised that I was walking/gliding beside her towards the opening in the wall. As we stepped into what appeared to be a dimly lit corridor I automatically turned with her towards the left, taking a few more steps, and then turning left again I went through another opening.

Larger in size to where we had come from, this room had square cut edges at every angle and was oblong in basic shape. I would have estimated it to be at least 30 feet in length, 15 feet wide and probably seven to eight feet in height. Three of its walls were completely featureless, whilst the fourth was covered from floor to ceiling and wall to wall in what appeared to be illuminated, oblong, glass fish tanks. Each tank was approximately three foot wide by 18 inches in height. Their interior glistened as if filled with a sparkling, electric-blue, incandescent substance similar to water, but which had some sort of jelly-like consistency that made it look as if it might be congealed. However, the most eccentric aspect about these tanks was that each and every one of them had, suspended within it yet without any obvious support, some sort of living, but barely animated, human-like creature.

I moved forward to look at these creatures more closely. Peering into the blue-ish looking liquid, I could see that they seemed to be floating. There must have been at least 20 to 25 of these tanks, all stacked neatly on top of each other; all subtlely illuminated by some unseen light source that only made the liquid sparkle even more. I remembered seeing a movie by Stanley Kubrick and Arthur C. Clarke called *2001: A Space Odyssey*. In the very last sequence of the film there is a 'star child' floating in a translucent bubble out in the middle of space – it looked like an unborn foetus, a human baby in a stage of development. (After recently researching this image, I would have to say that I saw a range of different sized foetuses in these tanks ranging from six to 12 weeks.)

The Witch made me aware telepathically that these human-like beings were very important; that they were 'good things'. I could sense that, although there was hardly any discernable movement, these life forms were very definitely alive. I was then 'told' that she was very proud of them and that I should be proud too. It wasn't made clear to me at the time why I should be proud, but I was aware that I was somehow connected to these beings on a biological level. I can only assume now that these foetuses were genetically produced from my own DNA.

There is constant speculation that the ETs are devoid of any sort of emotion and that they are attempting to source and acquire that ability from us through medical crossbreeding and associated procedures. Of course, there are some ETs – the small robotic/organic Greys come to mind – that appear to be very clinically cold and detached, a fair description although not completely true in my experience. The truth is that the ETs I have had contact with have a great depth of emotions but they are not ruled by them as we are. Based on my understanding, I can comfortably say that as The Witch and I stood silently observing these fascinating beings, our images reflected back from the apparent glass structure of the tank, I could sense her amusement. Without turning to look at her I felt a gentle ripple of energy pass between us that was generated by a clearly defined sense of humour.

She said to me, '*You call us aliens; but the last thing I am is alien.*'

I am sure that at that moment I knew, irrefutably, what she was trying to say: that from our very basic and simplistic human viewpoint *we* are related to *them*. But as time has passed, I am no longer sure if I have retained the more complex understanding of what she actually imparted. I can of course speculate, and other people who know of this experience have also tried to understand the meaning, but ultimately one thing I have learnt in my contact with these remarkable beings is that any assumption on our part with regard to our supposed understanding of their thoughts and behaviour cannot be considered a solid foundation on which to base our belief.

For me now the question of where they come from and who they are is overshadowed by what this contact means to the human race and what we should be doing about it. I don't believe the contact has ever been about *them* – on a superficial level maybe – but rather it is to do with who *we* are and how *we* wish to proceed as a race of sentient beings at this point in our evolution where the next decision we make and the next step that we take will be the most important in our history.

"The obscure we see eventually.
The completely obvious, it seems, takes longer."

Edward R. Murrow

Food For Thought: The Extended Family

The theory of evolution states that over time, as new species began to evolve, the more dominant the species the greater control it had and the more likely it would be able to reproduce and continue up the evolutionary ladder. Furthermore, characteristics from those dominant species would be the ones to carry through to future generations. Otherwise known as natural selection this gradual evolution accounts for the immensely diverse biological world we all live in today. In theory, evolution is believed to have taken billions and billions of years, which has frequently conflicted with the religious beliefs of many people throughout history. As most of us know by now Charles Darwin first recorded the concept of evolution in his 1859 book *The Origin of Species*, which offered the world of science its first rational and well-argued theory concerning the manner by which evolutionary global change had occurred and would continue to occur over time. And I pretty much subscribe to that theory except for one tiny point: when it comes to the evolutionary rise of Homo Sapiens, if we are descended directly from apes how is it that *they* are still here, apparently unchanged from how they have always fundamentally been? If they were our 'point of origin' surely they would have faded away as the redundant model. Although my general belief system with regards to natural evolution of the species is supportive of Mr. Darwin, I also subscribe to the whole God/Creation myth too, or, I should say, *god* with a small *g* and most definitely in the *plural*. I believe that we might be related, in a broad sense, to the ape, but if that is true that line of evolution was definitely broken at some point, or it might be better described as having been 'tinkered with'. I believe that our place on this planet and our true origin is far from simple. If we all came from the same mold – basic human being model number one – we would all look the same, wouldn't we? But we don't. From country-to-country, continent-to-continent, we all look different and it's not because of the climate and environment or any other old chestnut the anthropologists drag out to explain the quandary they find themselves in when attempting to explain away the obvious truth. We all look different from each other because I believe our origins are diverse; yes, we are all the same basic model – the bi-pedal human – but our *makers*, or *individual designers* might be a better way of describing them, came from different locations off planet. Therefore, it might be easier to view the Earth as a sort of showcase for life from throughout the cosmos. To coin a phrase, we are *multi-seeded*. I believe that life on other planets is nowhere near as diverse as we find it here – and by that I mean just the human beings. For instance, it is my belief that on planet X or Y the indigenous bi-peds, in every single location worldwide, will look exactly the same. The animal life might be more diverse than that but still not a patch on the planet that we call home.

So in short, this is how I see it:

- There are some species that are indigenous to this planet
- Various alien races throughout history have seeded the planet
- Some have crossbred with the native bi-pedal inhabitants
- Earth is unique in its diversity of species and environment

The work of the Human Genome Project, created in 1990 by the U.S. Department of Energy and U.S. National Institutes of Health to identify all the genes in the human DNA (the blueprint for the human body), is a vast and complex subject matter. Understanding what they do and how they do it is probably beyond the comprehension of most of us, but for the ones that do fully appreciate their work – good on ya! I will attempt however to explain – in the simplest terms – what they're up to, which will hopefully explain the importance of the bombshell they have discovered.

Since the creation of atomic bombs in the 1940s, scientists have tried to understand what radiation does to the human body and what possible mutations it can cause in human DNA. After nearly twenty years and the hard work of nearly 3,000 scientists around the world, there are some expected answers as well as some remarkable surprises. First of all, if one thinks of the human body as a book then all of the pages and text together are the genome (the *entirety* of an organism's hereditary information). The DNA in every cell contains the whole book about how to make a human body. The 'words' in each genome book are the genes (a unit of hereditary in a human body). The Human Genome Project expected to find that at least 100,000 genes were necessary to create a human body. But one big surprise is that the number of genes is only about 22,000, which is much less than several other Earth creatures, such as plants, some of which have an amazing 40,000 genes. The main reason for the smaller number of genes than expected is that human genes are capable of multitasking the production of proteins. It might be said that the human genome book has a more advanced text, but it still takes three billion letters to write out those 22,000 words. Currently the main problem in the project is this: what is the *punctuation* in a string of three billion letters? Scientists have discovered some sentence fragments, but still they do not know *full sentences*. It could still take quite a long time to work out how all of these 'letters' and 'words' make our bodies work. An even bigger puzzle is that inside some of the 22,000 genome 'words' are strings of repeating 'letters' – imagine seeing a hundred letters Xs together and not having any understanding of why they are there. Scientists call long strings of repeating letters 'non-coding sequences' or 'Junk DNA'. Most interestingly, there is an incredible amount of 'Junk DNA' in the human DNA molecule. In fact, that figure is an amazing 97%. It is suggested that this apparently 'useless' DNA has as yet an unknown function. Further it has been suggested by some scientists that our 'Junk DNA' is no less than the genetic code of an extraterrestrial life form, created by some kind of extraterrestrial programmer. In a nutshell, it is a working hypothesis within a progressive stream of scientists that a higher extraterrestrial life form was engaged in creating new life and planting it on the planet Earth. The 'Junk DNA' is therefore *their* fingerprint. Without any definite evidence on the part of these scientists to prove this theory though, as far as I can ascertain, it seems that this is yet another example of the Occam's Razor in operation. It does, however, fully support *my belief* on the matter. The hidden code in the human genome is still to be broken and the secreted message, reason and purpose still to be realised. So are *we* them or are *they* us? What a better way to leave us a message or calling card than within the very make-up of the human being. We continually ask to hear a message from the ETs; maybe it has already been left by our 'extended family'.

I'd like to leave you with one final thought on the subject: if an alien species has already been to this planet and 'tinkered' with our development, is it possible that this is part of an ongoing project, one that could still be operating today?

"Nothing in life is to be feared.
It is only to be understood."

Marie Curie

Experience: Wanstead Flats

East London – 1980

In the summer of 1978 I got married. I had met my future bride just 18 months before and without sufficient consideration to our extremely young age and relative immaturity we set sail on unchartered waters with no real map or compass for guidance. We were far too young to be 'playing house' and the reality of what that brings came as quite an emotional and cultural shock to us both – a pair of inexperienced, fledgling grown-ups. Although initially there was a novelty factor to the proceedings, once the honeymoon photos had been put away and we got down to the nitty-gritty of day-to-day life, it became apparent that for any young person their early twenties should be taken up with activity of a much different nature. What should have been an exciting voyage of discovery quickly lost its lustre as we were really just a pair of kids, unprepared for the responsibilities at hand.

One dark and frosty autumn evening in the year of 1980 my wife and I were travelling home amidst the rush hour traffic of east London. We were within a mile of our destination and as the line of cars finally ground to a halt our attention was simultaneously drawn to a light that hung low in the night sky just to our right. We crawled along in the slow moving traffic, still distracted by what we could now discern as a ball of light as it appeared to 'bob-along' just above the treetops. I wound the window down and watched as, whatever it was, came to a standstill. Although it was hard to make out against the backdrop of the night sky, the light did in fact appear to be attached to something: the faint outline was slim and oblong with the light towards one end. In a more pronounced movement this time, the light jumped from one position to another and then stopped abruptly once more.

To our left was a small two-storey block of residential apartments and the entrance to a private hospital just beyond. Further ahead and curving away to our right was a large area of grass and woodland known as Wanstead Flats. Hugging the edge of the road to the right was a line of trees and the light sat very close to the top of the trees, no more than fifty feet in the air. It was clear to me that I was observing something 'close by' and definitely not something at a distance.

I bumped the car partially up onto the pavement on my left and waited to get out as the frustrated and agitated drivers negotiated their way around my obstruction, all the time sounding their horns in obvious consternation. My wife and I gingerly stepped across the road, avoiding the cars coming from both ways all around us, and stood on the grass verge on the edge of the tree line. Although most of the trees had now shed their summer leaves it was still difficult to make out the light through the branches. We walked through the trees and onto the grassy common area beyond, where we could observe the light once more without hindrance. It was so incredibly low that it felt as if we could jump up and touch it. And then, as if a switch had been thrown, another light popped on towards the far end of the object, which effectively made it easier to make out the shape of this elusive phantom. It reminded me of the metallic tubes that expensive cigars came in, but was actually rounded at both ends. There was a ball of light at each end and it just sat there, hovering and occasionally oscillating as if caught in a current of air. I had the impression that it was probably no bigger than my car, either in length or depth.

I hadn't taken my eyes off this object since getting out of the car, and was aware that my wife was still by my side, her head tilted back and engaged in watching this peculiarity. It was then that I noticed I could no longer hear the meandering rumble of the traffic, although it was only a mere 20 or 30 feet behind me. Also, I was aware that I was only in shirt-sleeves having left my suit jacket hanging back in the car, but I did not feel ill at ease with the biting coldness of the night air. Although these thoughts were fleeting, they did momentarily register and made me wonder. Within just a few steps the object was now directly above us and I could clearly see the difference between it and the dark night sky. It was flatter somehow in colour, the consistency noticeably matt than the deep blackness beyond. And then the lights dimmed. In doing so I could discern the shape much better, it stood out, more pronounced somehow.

Something touched my left hand. I felt my wife's fingers interlock with mine. She squeezed my hand. I looked around and found her staring back at me, her eyes asking a rush of questions and yearning desperately for answers. I held her gaze for a brief moment and then looked back towards the object. The lights brightened once more and the object glided silently away, heading inwards towards the middle of the common and then stopped. I imagined it moved no more than 100 feet from where it had been, because within seconds we too had moved position and once more stood directly below it. If anything it now appeared lower. I felt that I could have picked up a stone or a branch and hit one of the lights dead centre.

As if a gust of wind had caught it, the object wobbled, then turned on its end with one light pointing to the sky and the other aimed down towards the ground. Slowly at first, but then picking up speed dramatically, it went straight up and within a matter of maybe five or six seconds it had completely disappeared from our sight.

I remember standing there for a couple of minutes at least, just looking up at the space where the object had been, but saying nothing. Eventually we made our way back to the car, noticing that the previously congested traffic had now dispersed and the road was relatively clear of other vehicles. Neither of us have very clear memories of our arrival home – although it was only two minutes down the road. All we have is a vague recollection of retiring quite early that night and of not actually discussing our encounter, either on that night or any other time since then.

As I have recounted the event here, the memory has returned in sharp and clear detail and I have been able to not only 'see and feel' each moment, but totally relive the experience as it transpired.

Although nearly 30 years ago, the memory of that evening was apparently perfectly recorded and remains entirely accessible and replayable in my mind.

"Why not go out on a limb?
Isn't that where the fruit is?"

Frank Skully – Author

Food For Thought: Tweaking

My interest in most things to do with 'outer-space' goes back to the late 1960s. As part of a global community that sat on the edge of their seats as Apollo 11 travelled to its place in history, I kept a wonderfully detailed scrap book of photos and reports cut daily from newspapers and magazines. I recall my fevered interest in anything and everything to do with that moon landing. It really felt as if things would never be the same again. There was a palpable energy of excitement that circled the globe in appreciation and wonderment at the achievement that we were all witnessing. Ironically, I can also remember everybody's interest begin to wane as the Apollo programme finally ground to a halt several years later. The marvel of putting men on the moon had not changed, just our perception of events. It happened years later too with the Space Shuttle; the first few launches were a matter of wonder and then they too became commonplace. That was until the day came when one of the missions ended in tragedy. Then the human spirit was reactivated and once again we paid attention. We lose touch with the wonder of things don't we? 'Life never changes,' is one of my favourite sayings, 'only how we respond to it'. There is only one reality though, but we have a billion perceptions of it, more often than not one perception fighting for supremacy over the other.

There is another moment that stands out in my memory from that time; when Eric Von Daniken's book Chariot of the Gods was released and serialised in a Sunday newspaper. The headline asked, 'Was God An Astronaut?' Before I read further, I already knew the answer was in the affirmative. What I read in the pages of this Sunday periodical felt so right; somehow I knew that what was being said was correct. I'm aware that Von Daniken has had his debunkers to deal with, but on the whole I believe he was correct in his basic hypothesis. I believe now, as I believed then, that our slow and gradual evolution was given somewhat of a boost through the intervention and genetic 'tweaking' by a race of beings superior to us.

It was such a time of wonderment for me as my prepubescent contact experience memories began to marry with the mature world of which I was starting to become a part.

"Where there is love there is life."

Mohandas Karamchand Gandhi

Experience: Little Star and Peter Pan
The Hybrid Babies

Location unknown – 1983

From my teenage years until the present day, my contact with Little Star has been unpredictable. Throughout this time, and just like everybody else in their life, there are certain memories that remain high in my conscious recollection and some that have faded with time due to their lack of incident. One might suggest that due to the exotic nature and location of my contact with Little Star I should be able to recall every single occasion that our paths have crossed and what exactly occurred. But it is not like that. Surprisingly when we are together there is a high degree of normality a lot of the time. In fact, the memory only tends to remain vital when our 'hosts' have clearly been present and when they have had an obvious input on the proceedings, although this is not always so. That is to say they are always present, though sometimes secreted 'behind the scenery'. There is one particular memory from my adult years, however, that stands out, which very clearly fits the above criteria.

To my recollection the location and environment for our meetings remains consistent: the room is domed in appearance, very much how one might imagine the interior structure and colour of an igloo, but on a much larger scale (and obviously not made of blocks of ice). The height is probably about 12 to 15 feet, whilst the diameter is maybe twice that size. The walls are bare of any decoration and there doesn't appear to be any machinery or equipment of any sort. When one finally takes the time to notice it is impossible to find how the room is lit, for the light source remains elusive. There is a sort of hazy quality to the radiance; although everything is always plain to see, it might be best described as diffused. The room temperature is neither hot nor cold, but occasionally there is a noticeable smell that defies description, only because I have never come across anything to compare it with. Concerning the noise now I come to really think about it, there has never been any sound as such, but I have been aware of a vibration at times, although this has just been 'in the air' rather than feeling movement through touching anything. Bizarrely, the openings into and out of the room tend to 'come and go'. I can imagine that must sound somewhat strange, but all I can say is that I have seen openings in the curved walls and then sometimes the walls are clear and smooth without interruption. On one occasion I could see a ramp that led from the surface of the floor and rose, hugging the wall, to a small horizontal platform about six feet off the ground. The ramp was unsupported, had no railings and appeared to protrude out of the wall. There wasn't any door or opening at the top of the ramp that I could see, but surprisingly given my curious nature I only remained puzzled momentarily. The floor of the room appeared to be of the same light-coloured material as the walls, which is quite cool to the touch and extremely smooth in texture. Having said that, I do recall a time when Little Star attempted to slide across the floor in her stocking feet but came to a complete standstill almost immediately.

On this particular occasion my shared contact experience with her started in the domed room. There was no conscious lead-up, no memory of being escorted from my home – my recall begins when I am with Little Star. I noticed how different she looked. I paid close attention to her body, which was now much older apparently having reached puberty, because there was a clear definition when it came to her breasts and hips (although her stature remained elfin). Her hair was now much longer: dark and silky and pulled back in a tight ponytail. She wore blue jeans and a dark colored, tight fitting t-shirt. Her feet were bare. I guessed she was probably about thirteen or fourteen years of age. It felt like I hadn't seen her for quite a while, although having said that, I immediately felt a closeness with her that never, ever seemed to dissipate, no matter what length of time interrupted our contact.

We were standing together, nonchalantly leaning against the wall, looking in on familiar surroundings. We were in the 'igloo room', but this time we were not alone. There in the middle of the room, lying on the floor was a group of very young, apparently ET babies, seven in total. They lay motionless on what appeared to be thick, oblong, rubbery-looking mats. They wore little cream-colored nightgowns that covered their whole bodies. Small bulbous hairless heads with translucent white skin protruded from the neck of each gown. Disproportionately large dark eyes stared out emotionless from each one. Each baby lay still and silent. It was impossible to detect which sex they were.

Standing by them were two groups of figures. One huddle was made up of four or five familiar beings – three or four foot tall in height with fragile looking bodies that seemed as if they would snap in a strong draught; shiny, pale grey, tight-fitting coveralls; huge hairless heads with large black eyes that wrapped from the centre of their expansive forehead to the side of their heads. They twitched and constantly turned their faces back and forth towards each other whilst emitting a barely audible clicking sound.

The other group was made up of four taller beings, maybe four feet in height, comparable in appearance to the smaller ones and not without a passing similarity to my regular guardian, The Witch. These beings wore a garment akin to the tiny babies. On the front of their outfit, covering the upper left breast was a symbol. It appeared to be a thin dark line with another line curled around it. I have since seen another emblem that reminded me of the one I saw: a caduceus, which depicts two snakes curled round a staff or a rod. It is a logo used in the medical profession and is sometimes referred to as the Staff of Asclepius or the Caduceus of Hermes. Additionally, I have also seen how DNA is sometimes depicted and this too reminds me of what I saw that day. As they moved I became aware that two of these beings were each holding a small baby in their arms. They walked over to where Little Star and I were standing and made it clear that they wanted us to take hold of the babies. I happily accepted the baby and held it, securely cradled, in my arms. I felt immediately disturbed by the way it looked. Obviously I'd seen young babies before and was always impressed, on the whole, by their healthy, robust appearance and their vibrant attentive eyes. This was not the case with this fragile creature who lay motionless in my arms. With my left hand I placed my palm directly onto its stomach and in doing so I experienced a tingling go through my fingers, quite like a pins-and-needles sensation. This appeared to have a direct effect on the baby, because it was then that it began to stir; moving its body ever so slightly and raising its arms from its side.

I noticed that Little Star, who was holding the other baby, had now sat cross-legged on the floor with it cradled in her lap. I followed suit, and in doing so it enabled me to free up both of my hands to attend to the tiny child in my care. A memory stirred that we had both done this before: at another time, in very similar circumstances, when we had cared for these strange looking children in this manner. It was then that my attention was broken. A voice in my mind asked, "Why do you want to do this?" I looked up to see one of the taller beings – Little Star referred to them as the Nurses – looking directly towards me. Although her mouth – what there was of it that is – did not move, I could still hear her voice. I didn't really understand the question and she seemed to respond to my unspoken thoughts. To my right, Little Star was laughing, and apparently aware of what I was hearing said, "That's obvious, isn't it?"

The Nurse now turned towards her. I heard the word, "Expand".

Little Star looked to me and said, "They're happy with us Peter, but they don't understand why we want to care for these babies. I told them it's obvious, but they don't get it." She laughed again.

The Nurses reached down and removed the babies from our care. Still on the floor, but now on our hands and knees, Little Star and I shuffled over to where the slightly older babies lay on the floor mats. Although a little larger than the other children, these ones still exhibited the same lacklustre composure and sickly appearance. I didn't think they were being fed enough. I engaged one of the Nurses and told her so. Her face remained blank and apparently emotionless but I sensed a reaction. But the same question just came back: "Why do you want to care for these children?" My response was plain and simple "It's the natural thing to do. It's what we do".

As I knelt over one of the babies, my head but a few inches from its face, the child, without any prior warning, became suddenly animated and reached up towards me with its little hands. Its spindly fingers instantly made contact with my cheeks; a mild pulse of electricity ran through my body galvanizing me rigid to the spot. It was as if the mind of the baby had entered mine. I was overwhelmed with an array of colours, flashing like a fluorescent rainbow. They spoke to me: not in words, but in emotions. I sensed a great wave of loneliness. Then came another of gratitude. Emotion filled my eyes and tears began to stream down my cheeks.

Slowly and carefully I broke free from the touch of this delicate and unpredictable child. I felt mildly nauseous and I thought I was actually going to be sick and I had a mild burning sensation in the centre of my chest. I began to sit upright and knelt back so that I rested on my heels. It was then that I felt a hand resting on the top of my head, its fingers curving onto my forehead. Believing it to be Little Star I turned my head slightly, surprised to find that it was one of the Nurses who was standing at my side. She looked attentively down at me; her large black eyes fixed on mine, drawing my mind in towards her. I felt a great sense of care and comfort coming from her and instinctively reached up and touched her fingers. I traced along the length of her fingers; they felt like long rounded tubes. I couldn't feel any knuckles. She bent towards me and placed another hand onto my chest. The fingers felt unnaturally long. Her touch seemed to relieve the feeling of sickness. A warmth spread from my solar plexus outwards through the whole of my body and I closed my eyes. I remained in that position, silent and still, for what seemed forever. When I finally opened my eyes again, the Nurse was no longer standing by me. I scanned the room to see that Little Star was pretty much in the same situation as me: kneeling down still, sitting back on her heels, eyes opened and staring blissfully, not really focused on anything in particular. She no longer held a baby. In fact, although the mats remained in the centre of the room, they were now unoccupied; the babies had gone. I looked around to see that we were quite alone. Somehow, whilst I had had my eyes closed the Nurses must have rounded up the children and left Little Star and myself alone in the room. How much time had passed, I had no idea.

What had been the purpose of what had happened, I wasn't really sure. Without a clear explanation of events, I could only wonder – then and now – of what it had all meant. What stands out for me is the question that was posed by the Nurse: why did we want to care for the children? The natural response, I suggest, in most human beings, is to show care and affection for a child, especially one under a year in age, and even more especially for newborns. It just seems to come naturally. But I suppose what the question really meant was why did Little Star and myself want to care for these particular babies – apparently with no direct link to ourselves and clearly quite 'alien' in appearance.

I did not believe then or now that the ETs did not understand or appreciate 'love' or 'care'. It has been suggested by numerous sources on many occasions that this is the definition of the scenario I found myself in and other incidences similar in nature to my own, but please do not fall in to that very obvious trap of assumption. The ETs might have a different perspective to us on a lot of things, but they totally understand what love and care is and how it works.

There is certainly something within the human psyche that enables us to step-up-to-the-plate and assist people in their hour of need: one only needs to look at times of obvious calamity, such as a war situation or catastrophes resulting from an earthquake or a terrorist attack. Increasingly in this modern world, especially with the aid of 24-hour worldwide news bulletins, we witness the likes of 9/11 and tsunamis rolling out before our eyes and seeing brave individuals coming to the rescue of fellow human beings. Where does the motivation for these selfless acts come from? The human race is truly at its most wonderful and spectacular when we are in the tightest corner, when the proverbial *merde* is hitting the fan and when the odds are against us.

There is a bottomless well of theories being bandied about on the Internet concerning the whole subject matter of contact with aliens – some have foundation for further discussion and some are just plain goofy. Sadly, very little verbal speculation and discussion, if any, comes from the most important people of all, those who have a right to 'have their say' and are probably more informed to do so – the Experiencers themselves. Because of the ridicule that is often directed their way, they tend to remain relatively quiet, which is so very poignant considering their potential to help bring clarity and focus to a continuing blurry image. The so-called UFO 'experts' and self-absorbed scientific community continue to ignore, comparatively speaking, these very important individuals and remain content to regurgitate a series of dead-end hypotheses that enable them to remain in print, in the spotlight and on the lecture circuit. I have often wondered why these apparently smart and savvy people are not approaching the Experiencers and trawling the depths of their experiences for the proverbial Rosetta Stone that might prove to be the key to this unexplained phenomenon.

As a member of this ever-growing league of frontline human-guinea-pig Experiencers, I would like to emphasize that this kindly suggestion does not come from a place of ego. I truthfully and gratefully acknowledge the real proficiency and merit of the 'experts' of which I speak, but it's just a shame they appear to be stuck in some sort of intellectual maze that keeps them from taking a leap of faith, one that would ultimately add them to a list of brave individuals willing to break free from the safety of the 'known universe' and expose the breathtaking magnificence of the 'unknown'.

Intermittently peppered within the overall maze of conspiracies is the subject of inter-breeding between humans and aliens. Amongst the diverse dialogue that goes on you will find it suggested that aliens are abducting some humans to extract relevant physical materials to aid them in their creation of a hybrid human/alien race of beings. This may be possible of course, and even I have had an experience to suggest that I could have played a part in a programme of this nature. Additionally, many authors have explored this theory and documented their findings in numerous books and articles over the years. Their thoughts range from a) the ridiculous: speculative paranoia dredged up from the muddied waters of insecurity created from the unseen manipulation of those that would benefit from our ongoing fears; and b) insights drawn from a more solid foundation of intellectual consideration based on what is actually being presented, but still sadly confined within the limiting and restrictive parameters of current human thinking. There is, however, an ever growing paradigm of thought that is viewing our world from a different perspective, one that is not retarded by the redundant and crippling belief that the world is still flat in more ways than one. The fathers of this pioneering wave – John A. Keel, Jacques Valle, John Spencer and Dr. John E. Mack – and their documented work on the subject, is continuing to influence the next generation of like-minds who refuse to remain silent at a time when their voices need to be heard at their loudest and most profound.

We all need to be tuning into the more obscure bands of transmission, because it is within those explorations where we will discover the road to enlightenment. The truth – for there is only one – as *The X-Files'* Fox Mulder once told us, is most definitely out there. And, I believe that the truth is obvious, that we are somehow looking *through it* or just being distracted by the smoke and mirrors of the conjurer. Maybe the truth is not *out there*, but actually *in here*: in all of us.

"When the mind is thinking, it is talking to itself."

Plato, Greek philosopher and mathematician

Food For Thought: Communication

I am regularly asked how do the ETs communicate with me, as well as how do I hear and respond to them. My ability to articulate something that normally resists explanation because of its ethereal quality has grown for two main reasons: first of all because of the repetitive contact experiences; and secondly, and far more importantly and relevant, because of something that 'they' have done to me. On two specific and separate occasions, when in contact with two very different types of ETs – the 'uniform' and 'factory-processed' looking species, commonly known as The Greys and the apparently and very possibly more highly-developed race referred to as The Light Beings (which I will describe in greater detail in a later part of my story) – I have had procedures carried out that have altered certain aspects of my being and behaviour.

It seems to me that whilst outwardly uniform to a degree The Greys do indeed have some sort of hierarchy be it based on their abilities or their own intellectual and spiritual development or maybe even organic structure and design. I have had regular contact with an advanced member of their race, whom I refer to as The Witch. On one particular occasion a change occurred with regard to how I saw my role in the contact experience, shifting from one of disempowered abductee/victim to one of empowered experiencer/participator. Apparently because of this 'show of strength', I believe that a decision was taken involving a placement or replacement of some sort of mechanical/organic device in my body that affected me on a physical and behavioural level.

First of all, and very noticeable, was an instant inability to consume and digest any sort of alcohol or animal flesh without becoming physically sick. I then finally had complete and unrestricted access (after a gradual transition since my 'awakening' of 2001) to contact experience memories that had previously been partly out of reach to my conscious mind. I began to experience an extended capacity and depth to my senses, emotions and feelings on a normal physical level, but also within an ever-increasing awareness to my metaphysical range of senses. At first, I did not deal with it at all well and my body responded negatively to the new demands that were being made of it. Clearly, as I was to find out, there was a period of adaptation required. Gradually this has occurred, although there are still times when it remains a challenge to my physical body, perhaps because of its present innate design. Therefore, I am afraid to say that on a physical level my body does 'complain' due to the things I have gone through in my contact experiences, but this particular subject is for later on. Additionally, the non-physical changes that occurred to me have also been an ongoing trial, after having to review and refine a lifetime of habits and beliefs. Everything, I have found, improves with practice, even the evolvement of ones own spiritual enlightenment!

Then, since my contact experience with The Light Beings, after an incident in 2004, and progressively since that time, so many other aspects of my contact shifted into a higher gear. For instance, it feels now that I have some sort of 'open line' when it comes to their contact with me and my ability to ask and receive answers. Although difficult to explain in simple terms, what it boils down to is that I only have to consciously focus and amplify my intent to 'speak' with them and I immediately experience a clear and defined response. Additionally, totally out of the blue, but with similar definition 'they' appear in my conscious mind with an understandable dialogue. Which brings us to the million-dollar question: how exactly does that work?

I have heard other Experiencers talk of a type of telepathy occurring where they hear a voice inside their head but notice no apparent projection or application of sound coming from the ET involved. They have also spoken of not articulating their own thoughts 'out loud' but merely just having to 'think' something and it is clearly understood and responded to.

Well that's how it initially started out for me too. In my early contact and throughout most of my teen and early adult experiences, telepathy was, as far as I can actually recall, how the communication between the ETs and myself occurred. Then it all changed.

Since my contact with The Light Beings began, the way I 'receive' has changed, although the way I 'project' does not appear to have altered; on an unexplainable level, however, I know for definite that it has. I believe that all I still do is 'think' something and 'they' understand it, but I sense that this is simply my human view of what is happening. Let me explain about my 'reception' first and then you may be able to appreciate what I mean. When 'they' communicate with me, it is with symbols, memory, emotion, stimulation of the five senses, metaphor – a seemingly never ever ending list of feelings, pictures and experiences. Therefore, when I 'talk back' I believe that I am just communicating with my unspoken words, but I have a sense that there is some sort of implanted organic/mechanical device that is translating it all for me into a more intricate form of language or that 'they' have activated a long-forgotten tool that resides in all of us, enabling us to be able to transcend our basic form of language. I have a sense it is the latter, which means we are all capable of doing what I do and probably will in the near future.

"Every truth passes through three stages before it is recognised.
In the first it is ridiculed,
in the second it is opposed,
in the third it is regarded as self-evident."

Arthur Schopenhauer

Experience: Thunder Road

London suburbs – February 1986

In 1985, a lucky shooting star fell directly into my lap. At the time I was working in the newspaper industry and I had been approached by the managing director of the group I worked for to create my own independent company to distribute a number of our newspaper titles to points of sale within the North London and Hertfordshire area. I had previously been part of the in-house distribution system, but now the senior management wanted the job carried out by a third party, thus alleviating themselves from the management and administration of the work. Although this might have seemed an unusual practice, the fact that the distribution would take place at night and would in no way impact on my daytime activities, meant that I eagerly went ahead and met the requirement of the newspaper. In fact, I always felt that I had a vested interest to make sure that the job was always done to the highest standards.

The company that I created was called 'Thunder Road Distribution'. Being a fan of Bruce Springsteen, a well-known song of his – *Thunder Road* – became a perfect title to dovetail with the nature of the work. The distribution was carried out late on Wednesday night and through to the early hours of the next morning. I enjoyed certain aspects of the work, especially the clear open roads that I navigated without the hassle of normal daytime congestion. To keep me company throughout the process of delivery I would listen to music tapes or night-time radio programmes. Whilst my fellow delivery drivers were delivering in inner London, the major part of my particular delivery route took me off the main road system and onto country lanes that wound their way though rural Hertfordshire. On this part of the journey the road lights completely disappeared and I normally had to turn on the full beam of the headlights to light the way. I would often slow down and very quickly turn my lights on and off, and by doing so would get a quick glimpse of the night sky which was, although not perfect, relatively unaffected by light pollution from the city. I had a fascination with the night sky and loved to watch the myriad twinkling stars above me. As time went on I began to leave the lights off for longer and longer as my vehicle slowly crawled along the lane. In a game of 'chicken' I would wait until the very last second before turning the lights back on. The rush of adrenaline swept through my body each time as my focus bounced back from the darkened lane to the jewel-bedecked night sky. Knowing that at some time my luck would eventually run out and that I would probably plough my delivery van into an unseen ditch, I decided to pull over and park whilst making my nocturnal observations.

On this particular night in February 1986, just before my 29th birthday, as the van came to a standstill I had already turned off the headlights and was sticking my head out of the side window to look up. The cold air took my breath away, but within seconds I knew I was going to introduce a new element to my sky watching: I was going to get out of the vehicle.

Although it was absolutely freezing, I was wrapped up very warmly with the added bonus of thermal underwear beneath my normal delivery outfit. With a wool hat pulled down snugly over my ears, I stepped gingerly from the van and walked round to the front with the intention of sitting on the hood, which was still warm from the heat of the engine.

The comforting warmth finally made its way through the clothing layers and I sat with my knees tucked underneath my chin and my arms wrapped around my legs. Struggling, as usual, to take in the enormity of the majestic night sky I sat transfixed with my eyes to the heavens. I tried telling myself that I was seeing suns, not stars, and that around each sun could possibly be a planet just like Earth. I toyed with the idea that on one of those planets there could be a man sitting on the hood of his van looking back at me. However hard I tried, even though I was able to occasionally connect with the concept, it slipped away like grains of sand through my fingers. It wasn't as if I didn't believe in life on other planets, quite the contrary – I knew differently after all – but how could I ever reach out and make contact with that other person? It simply blew my mind.

The timing of the job always ran like clockwork: I would finish delivering by 2.30 to 3.00 a.m. each week, return the hire vehicle to the garage lock-up and then be back home by 3.30 a.m. I knew that I couldn't sit there all night, so after about ten minutes of stargazing I looked at my watch to see it was now 2.50 a.m. and so I started to get down in preparation to return to the cosy interior of the van. Reluctantly searching in my pocket for the ignition keys I realised that they must still be in the ignition. It was then that I felt the air pressure around me change. It was as if the air had been suddenly sucked away from all around me. It made me catch my breath. My ears popped as they do on an aeroplane.

Steadying myself with my left hand on the hood of the van, I peered into the darkness of the lane as it indistinctly stretched out in front of me. Somewhere, just beyond my ability to clearly define detail, I could see a glow. About the size of a football and roughly a few hundred feet from me, the muted light source was hovering around head height. As if mesmerised by what I was seeing, I stood frozen as I became aware of a faint and soft rumble. It wasn't felt through the ground, but rather the slight tremor was in the air.

I stared intently at the distant glow, my eyes fixed keenly on its centre. It seemed, at first, as if it began to move towards me, but I quickly realised that this was not so. Whilst the glow somehow remained stationary, the area around it appeared to stretch in my direction. It was totally perplexing to watch as the light grew larger, but I knew instinctively that it wasn't moving, as I understood it at least. At the same time I could feel pressure in the area of my solar plexus. In a similar but still unexplainable manner the pressure was apparently stretching the space that I stood in towards the light. My chest moved out as my body was pulled forward by an unseen force.

I watched as my surroundings distorted: trees expanded, bushes twisted, grass verges warped before me. A sickening feeling overcame me as my mind swam with a million confusing images: bodies paraded about me with swift and determined action; lights pulsed and flashed in my face; an unrecognisable torrent of words invaded my head. All the time whilst my feet appeared to remain fixed firmly to the ground, a sense of movement continued to pervade my whole being. And then there was nothing.

The illuminated motorway sign that was filling the expanse of my car windscreen indicated that the next turning was mine. My right foot instinctively rose to apply pressure to the brake pedal and my hand moved to pull down the indicator. As the car slowed and moved from the inside lane to the slip road I felt confused. I was on my way home after working all night? Yes. It seemed hard to recall the moment I had returned the van and picked my car up to make the journey along the motorway. I felt very sleepy and just a little disorientated. Maybe that was it; I was just overtired. Somehow or other, I was on 'automatic pilot'.

Driving through Chingford in east London towards my home I noticed the interior of several shops alight. There was movement through the windows as shopkeepers prepared for another working day. Instantly, I knew that there was something very wrong with what I was seeing. Although it could have only been around 3.30 a.m., there appeared to be businesses beginning to open their doors. Slowing down to look more closely I could see that it wasn't all of the stores, just certain ones: newsagents, bakers, grocers. Although these were establishments that always opened earlier than others, this was far too early.

Street lights were beginning to go off with no apparent affect on my ability to see. Although the headlights of my car continued to illuminate the road, my surroundings to the left and right of me appeared to be bright enough to see without the aid of artificial lighting. I looked to the sky and saw the first faint suggestion of dawn as it began to brighten the sky and alter the overall ambience of everything around me. I depressed the gas pedal and continued on my way home.

By the time I reached my house, there wasn't a streetlight left on. The sky was quite bright. It was clearly morning. I then did what I should have done much sooner, I looked at my watch. To my utter astonishment I saw that it was apparently just before 7.25 a.m. Although confused beyond belief, I instantly calculated that I was, in fact, four hours late!

As I pulled up outside my terraced house I could see that there were lights on downstairs. The front door opened before I had a chance to get my key in the lock. I looked at the face of my wife that said just one thing: "Where the hell have you been?" As hours of apparent worry drained from her face in a moment, tears flooded down her cheeks, she reached out and wrapped her arms around me. Within seconds her mood had swung one hundred and eighty degrees and she was pushing me away. Immediately the questions started. However which way my wife asked me why I was so late, I was unable to give her an answer. After several more attempts to rephrase what was, after all, clearly a very simple question, my wife gave up because of my lack of response. As I remained in the hallway bewildered and confused, she stomped up the stairs and slammed the bedroom door behind her.

Because of the late hour, I bathed and got ready for my day job. There would be no time for a catch-up nap before going out this time. I left my wife in bed and left the house once more. I didn't return again until after 5 o'clock that evening. It came as no surprise that my arrival was greeted by a somewhat frosty response. Although my wife was normally very good at 'letting things go' there was clearly to be a more lengthy transition this time. As I was still extremely tired because I had not slept since Tuesday night, I had no intention of arguing or even attempting to explore or explain my unexplainable circumstances. So I just left it alone. My wife and I shared an evening meal, watched some television and retired at a very early hour. The matter was just left; we never spoke of it again.

I knew that some sort of contact experience must have occurred, but as usual my subconscious mind very quickly placed it into the 'coping' compartment of my brain, thus allowing me to move on and proceed with the day-to-day elements of my normal world. I simply let it go. Why 'they' did not want me to know what had happened or forgot to reactivate the memory I will never know. Even after my profound reawakening of 2001, it still can happen that way. And I trust 'them' in their judgment. As far as I am concerned it matters not either way. I have so much conscious recall of my contact experiences now, that the occasional event that slips through the net must do so for a reason, so I am not going to waste my time dredging my memory banks anymore for windows of experience that remain firmly closed. Que sera, say I!

"Memory is the treasury and guardian of all things."

Marcus Tullius Cicero, Roman philosopher

Food For Thought: Memory

Part 2: Does Our Heart Have Memory?

I suppose that once in a while we all give a thought to our memory (more so as we get older I am finding). We might, for instance, speculate how well it is compared to other people we know. In recent years, there has been a lot of research carried out to determine the capacity and limitations of the human memory. Beyond any doubt though, the human memory is one of the most amazing mysteries known to man.

We have already speculated about how the memory might work and the parts of the brain it is associated with, notably the frontal lobe, the temporal lobe, the cortex, and the hippocampus. These disparate parts of the brain work together, thereby allowing us to formulate and store information in the form of memories. Since human memory is such a complex system, it cannot be nailed down to an exact science, but it does appear that no one single area stores or processes information. In fact, it might be safe to say that memories are 'scattered' all over various regions of the brain. Taking this concept one step forward, should we confine the idea of storage purely to the parameters of the physical brain?

On 2nd June 2005 the US broadcasting company NBC reported a story concerning Christina Santhouse aged eight from Philadelphia, who had caught a virus that caused a rare brain disorder known as Rasmussen's Syndrome. Her doctor had performed a very serious and intricate operation called a hemispherectomy, a surgical procedure where one cerebral hemisphere (half of the brain) is removed or disabled. After the surgery, she had a slight limp and her left hand didn't work at all. She also lost her peripheral vision, but otherwise, she was an ordinary teenager who ten years later was able to graduate from high school with honors. There was a similar case reported in the *Daily Telegraph* newspaper in the UK on 29th May 2002: a girl named Bursa had the same disorder and her left brain was removed when she was three years of age. She went on to become fluent in Dutch and Turkish when she was seven years old. In another report it is stated that in 1987 a patient who had gone through a hemispherectomy had completed college, attended graduate school and scored above average on intelligence tests. In fact, studies have found no significant long-term effects on memory, personality, or humour after the procedure, and minimal changes in cognitive function overall.

The outcome of hemispherectomy is surprising. Neuroscience tends to suggest that memory is stored in the neurons of the brain. If that premise stands true and memory is stored in the network structure of neurons as one school of cognitive physiology suggests removing half the brain would destroy one's memory. Alternatively, if bits of memory information are stored in individual neurons in the brain as suggested by another school of cognitive neuroscience removing half the brain would at the very least destroy half of the memory. But it is apparent that some results disagree with both of the explanations. Removing part of the brain has been one of the standard surgical operations for severe epilepsy and has been performed thousands of times.

The orthodox explanation for this is that information stored in the infected brain areas is duplicated in the healthy part of the brain at some point prior to the surgery being undertaken, which is, I'm sure you'll agree, a strange concept but not that peculiar if we accept that the brain carries out some sort of re-allocation of memory due to the infected area shutting down. This rationalization is still inadequate when you take into account how brain surgery is performed. The surgeon has to remove the infected area and some surrounding healthy tissue – sometimes a much larger tissue area than the infected part – to make sure all infection has been removed. If the information stored in the infected area is reproduced somewhere in the brain before surgical procedure, it stands to reason that some information must be lost when surrounding healthy brain tissue is removed, thereby adversely affecting the memory.

However, this is not observed after surgery, so it is necessary to assume that the memory stored in the neighbouring healthy tissue is also replicated in other parts of the brain. This raises a question: how does the brain know how much healthy tissue is going to be taken out? If the brain does not know, surely then surgery will inevitably destroy or cause the destruction of part of the memory. The belief that memory is stored in the brain (in neurons or in the network of neurons) apparently contradicts the findings of brain surgery.

It is believed that instincts are inherited yet nobody has any idea where this information is stored. Cognitive memory is thought to be acquired through experience and stored by changing the signal chemicals in the neurons in the brain. Therefore, the late 19th/early 20th century American philosopher William James held that consciousness operates through the brain rather than the brain producing consciousness. The notion that consciousness is separated from the body has a long tradition in western thinking. Plato portrayed the earthly body as a limiting factor on conscious experience. Immanuel Kant, the 18th century German philosopher, put forward the body as 'an imposition to our pure spiritual life'. This idea matured into a proposition called Transmission Hypothesis, whereby the brain and body serve not as the originators of consciousness but rather as its trans-receiver. The cited supporting evidence for this hypothesis are mostly in the typically considered 'unscientific fields', such as psychedelic research, PSI effect*, after-death experience, etc. As a result, this hypothesis has been ill received within the philosophic and scientific community, but that is beginning to change based on progressive new research. Therefore, the idea of separating consciousness from the body might be a very sensible thing to do in the light of the above facts. If memory does not reside inside the brain, the functions of the brain need to be reinvestigated. It is possible that the brain acts as a bridge to consciousness. The similarity between the two is obvious, and the brain is the only pathway to consciousness and memory for both cases. The importance of brain to memory has been supported by a vast number of critical researches over a long history. But concrete evidence to suggest that the brain is the only organ associated with memory is lacking, on the contrary, some evidence suggest that the heart might be associated with memory too. Does the heart have memory? The question has been around for years. The question arose from years of transplanting the heart or other organs into human beings and then noticing some personality changes in the recipients. For some there is an overwhelming need to consume quantities of Mexican foods when that type of cuisine was never a favourite. For others, a sudden love for football, when sports were previously hated, comes into play. How can this phenomenon be explained? Can the heart actually feel, think, and remember? The answer could shed light on how memory is handled by humans. Rollin MacCraty from California has devised tests that show how the heart processes information. His tests show that when encountering an emotional event the heart responded before the brain. He duly concluded that the heart must have the ability to process emotional data. To associate heart with memory is a legitimate proposition based on these findings, but there is no medical evidence indicating that changing the heart to a mechanical heart leads to memory loss. This implies that memory is not stored in the heart. Could it be again, that like the brain the heart does not store memory but is a *gateway* to the memory? What kind of memory can be accessed through the heart? Are other organs gateways to limited memory too? These questions ask for the expansion of memory research to a much wider area besides the brain.

To be concluded in:
  Food For Thought: Memory
  Part 3: Non-Observable Spatial Dimension

"It is in our dreams that the submerged truth
sometimes comes to the top."

Virginia Woolf

Experience: Betty

London – Spring 1991

Before I met and married my new wife, Annie, she used to teach a weekly psychic development class from her home in North Chingford in east London. That's how I first met her, by becoming one of her students. The content of the course was based upon the spiritual teachings of a remarkable woman called Betty Balcombe. Annie had studied with Betty for many years and with her blessing and encouragement was now proficient enough to conduct Betty's teaching programme herself.

Betty still held group-teaching classes and also one-to-one consultations where she would respond to specific questions posed by the individual and offer support and guidance through what might now be termed life-coaching. Betty's technique, however, was very much metaphysical in nature and did not in anyway resemble the style and content of today's coaches, who offer clinical and scientific nurture to their client – Betty has always worked very much from the heart and from the soul. The description of the content and style of Betty's teachings as explained by me will sadly fall short of the actual experience, for I feel that it is one of those 'Grand Canyon moments' where you 'have to be there' to appreciate and understand. An alternative to this would be to read the two books that she has written, *As I See It: Psychic's Guide to Developing Your Sensing and Healing Abilities* and *The Energy Connection: Answers to Life's*

Important Questions – both of these books written by Betty are still available through Amazon.co.uk. However, if I were pressed to give a thumbnail impression of her work, I would have to say this:

Betty always said that we were here to take responsibility for our words and actions; to respond to every living thing with care and sensitivity; and to show love and guardianship for our home, the planet Earth.

Simplistic in nature? *Of course.* Achievable and realistic? *Most definitely yes.* The content might even sound familiar, you might say? There definitely have been numerous people throughout human history that have conveyed the same message, but the pure meaning ultimately gets entangled and lost within the sticky and manipulative spider web of religion or the power games of those that would seek to control us all.

The right way to live our lives properly, after all, has never changed; it just appears to take a long time to sink in though. As this innate wisdom echoes down through the ages and is once more heard by those who choose to hear, we must take ownership of the true meaning contained within and to ensure that we begin to live our lives with care and responsibility. For many reasons it will not be easy. For one thing, we will have to replace our knee-jerk, fight-or-flight survival instincts with responses of a more sensitive nature. When human beings slow down enough to consider their actions properly, it is easier to live responsibly, but that is not how we have been programmed to respond: our reaction clock runs much too fast and is too rigid!

Back in 1991, Annie suggested that before I began her class it might be advantageous for me to meet Betty for a one-to-one consultation. Betty's history and lineage is really quite 'exotic' and is not for me to divulge at this time but suffice to say I was drawn to her teachings for more than the obvious reasons of expanding my self-awareness and self-development. Who I am, why I am here and my 'contact experiences' are very closely linked with Betty on many different levels and it is very clear to me now why I was attracted to and drawn into her orbit. I hate to appear so ambiguous at this stage – especially after the honest and comprehensive telling of my story thus far – but I am certain that on a more subtle level you will be able to ascertain the true veracity of my contact and relationship with this remarkable individual without me breaking the integrity of our more intimate dialogues as I am unable to obtain her permission, for sadly Betty is now unable to furnish me with her consent on the matter, but not for the reason you may think.

At the time Betty was living with her husband in a very ordinary detached house in Surrey. This was where she held her weekly group classes and saw people on an individual basis. Annie had given me her address and the night before I was due for my appointment I retired to bed excited at the prospect of finally meeting Betty. I still have perfect recall of the 'dream' I had that night.

I dreamt that I met Betty at her home, although how she appeared to me in my dream turned out to be quite different from how she actually looks, although I was to find out later that it was interestingly similar to how Betty looks when other people dream about her. Additionally, the interior of her dream house was radically different from the non-descript, 1930s, suburban domicile that I actually visited.

In the dream I remember walking into a huge room with a high ceiling and ornate cornice. The only piece of furniture was an unusually high-backed chair, draped in a dark green throw upon which Betty sat majestically. She appeared to be very large in stature with slender arms and willowy fingers that were entwined in her lap. Her face was pale like polished ivory with pronounced cheekbones, mesmerizing dark eyes and long, straight red hair that perfectly framed her refined features and fell down over her shoulders. She wore a plain dark velvet dress that flowed uninterrupted to her ankles with sleeves that billowed out to her beautifully sculptured hands. Behind where she sat and just to the right the entire wall (or more possibly a window) was covered with a sumptuous red curtain that was suspended elegantly from a gold-coloured pole and hung in heavy pleats to the floor.

Betty gestured with a hand and I sat, very self consciously, on the floor by her feet. Not really knowing what to say, I just sat there and didn't speak. She lent forward and stroked one of my cheeks with a curled-up finger. Her touch felt warm and I instantly felt at ease.

"I understand you're coming to see me tomorrow and I thought it might be nice just to make your acquaintance before you arrived," Betty said in a soft, whispered tone.

I smiled. Her eyes held mine; they were so dark, black even, surrounded by whites of snow. She raised her left arm and gestured towards the red curtain over her shoulder. As she did so the sleeve of her dress wafted from side to side, as if in slow motion, which momentarily held my gaze as if in a trance.

"You may leave through there and I will look forward to seeing you again very soon."

I got to my feet and walked towards the curtain, all the time watching her eyes as they traced my movement. Moving my fingers along the folds of the curtain I found that there were, in fact, two curtains with a gap in-between. Pulling them apart I was surprised to find not a window but just an opening in the wall – like a regular sized doorway without a door – and sufficiently large enough to walk through. I hesitantly moved forward and found myself looking down into another room, for I appeared to be high up and peering from about seven to eight feet onto a double-sized bed that lay outstretched below. And to my even greater astonishment there lay two people – a woman and a man – apparently fast asleep beneath the flower-patterned bed covers. The room was in fact my bedroom at home and the people were my wife and myself. Before I knew it I was stepping through the opening and tumbling down towards the bed, heading directly for the sleeping me. Surprisingly what should have taken but a second, seemed to last forever as I floated down like a feather. Just before I finally made contact, my perspective altered: I was no longer looking down, but was now looking up, as if I were already prostrate on the bed and observing the other me as it fell from the opening above. And then there was a breathtaking shudder as the other body made contact with mine... and we appeared to become one.

I lay there wide-awake with my hands turned palm down feeling the smooth, soft sheet beneath me. I can clearly recall that sensation. After being in what had amounted to a dream sequence and then experiencing the break through into the real world, I can remember the vibrant transition that occurred. It wasn't smooth. It wasn't unpleasant, but it wasn't smooth. I felt my spirit body move from one dimension to another and merge once more with my physical body.

I could feel the breath in my chest rise and fall. With my available senses I took in my surroundings, attempting to confirm the reality of my situation.

I attempted to somehow process what had just happened, but in that attempt my mind began to unavoidably drift – just like it does in that state between wakefulness and sleep – and before I knew it my consciousness had slipped and I was asleep.

What had I experienced? Had it been just a normal dream? A sequence of thoughts that had occurred and been experienced within the electrical interactions of my physical brain perhaps? Or had my spirit body been summoned to another location where a meeting of sorts had been played out? Was this in fact a normal experience and one that is more often than not forgotten because of the apparently in-built software that files these events away in an inaccessible folder within the human memory system? Whatever had happened, it was important enough for me to remember the dream clearly the morning after and nearly 19 years later. I certainly remember the event differently to how I have remembered normal dreams. Actually, I remember it as if it really happened; the same way I remember where I went on holiday last year, what my son's name is or what I had for breakfast this morning.

I'm sure we can all identify the relevant milestones in our lives that have turned out to have had a profound impact on our ongoing development and the path our life ultimately takes. They normally revolve around specific events, special people and memorable decisions. We can normally remember, with great clarity, the moments when these things occur. They stand out for many reasons and often because there is sometimes no turning back after they happen. Hopefully they are remembered for a good reason, but that is not always the case. One thing that I have garnered from the more challenging changes that have occurred in my life is that there is wisdom to be gained from all events. More often than not that is why the universe puts forward these opportunities in the first place. And it has always been the same hasn't it, for all of us, throughout the history of mankind. Situations don't alter, challenges never change, but our response to them can, based on the wisdom we have acquired up until that moment.

I knew then and I definitely know now, that meeting Annie, which led on to being introduced to Betty and her teachings, was not a mere coincidence, not just a simple chain of events, but was very much a planned episode in my life. Planned at what time, by whom and under what circumstances is not relevant, for that is how these things work in how we live our lives. What is important is that we recognize these moments when they present themselves and respond in the appropriate way to enrich our life journey; for we must never forget that we are beings of free will and always have the opportunity to side step these chances if we so choose to do.

The events of which I have spoken in this section are intrinsically interwoven with my ET contact, as are all the other milestone events of my life. My ET contact remains a major thread in my time here and I acknowledge it to be a learning tool designed to aid and promote my growth as a human being.

There are countless other milestone events that stand out – no greater thing than the birth of my son – but there are others that have been infinitely more subtle. People have entered my world; thoughts, feelings and experiences have been shared and I have responded accordingly. There is wisdom to be had in the finer detail. I wonder where I would be now if the finer detail were to be altered in some way. I'm sure that on an alternative road somewhere those realities are being lived out. For are we not learning from the progressive thinkers in the world of Quantum Physics that all realities are possible?

So it is from this perspective that I truly believe that the human race is capable of change, a change so breathtakingly profound that it would fulfill the greatest hopes and dreams of mankind. Just consider that our world and how we live our lives could be unrecognizably different by tomorrow if we were to think with care and behave with awareness and responsibility for our words and actions. If we were to lay down our proverbial weapons of *protection/destruction*, life would be very different from how we currently experience it. An alternative quantum reality is just waiting for us to take ownership.

"Aerodynamically, the bumble-bee shouldn't be able to fly,
but the bumble-bee doesn't know it so it goes on flying anyway."

Mary Kay Ash

Experience: The Mandarin Oriental Hotel

Hong Kong – September 1999

In September of 1999 Annie and I flew off to Galway on the west coast of southern Ireland. Annie's two children, Charlie and Rachel came with us, as did my son Reece. It was an extremely special occasion as Annie and I were *pledging our troth*. It was our second time 'round-the-block' for both of us and we wanted a very quiet family only affair. Also in attendance on that day were our dear friend Karen and her daughter Millie, who are Irish and live in Galway. We had chosen this particular geographical location for our ceremony because of our passion for the west coast of Ireland and very specifically for Galway city, which held so many glorious memories from previous trips to the Emerald Isle. Although the weather was being true to form, with the wind lashing off the Galway Bay and projecting the customary rainfall horizontally into our faces, it could not dampen the romantic blanket that securely wrapped us both within its folds.

The ceremony itself was rather amusing because the facilitator was conducting a new type of service for the very first time and had to constantly check her notes to ensure that she had covered each and every part of the proceedings. My twelve-year-old son, rigged out in his smart Ben Sherman shirt and neatly pressed black trousers, handed over the ring in his capacity as my mini Best Man and the ceremony drew to a close with the cherubic voice of Bryan Kennedy singing from the small, portable CD player that we had brought along for that purpose. Retiring to a local restaurant for the wedding celebrations, a table had been specially and glasses were charged and toasts made. My new wife sat beside me looking more radiant than I had ever seen her before in an ivory-coloured silk dress and her hair of 'ever changing hues' cascading all about her face in soft, fluffy curls.

Upon our return to England we just had time to re-pack our suitcases before we made our way to Heathrow Airport to begin our long-haul flight to our honeymoon destination of Bali. We had arranged to break our journey into stages with an overnight stop-off in the exotic city of Hong Kong. So after a rather lengthy and tiring twelve-hour flight, we disembarked our Cathay Pacific aeroplane and were shuttled from the airport to our beautiful five-star hotel, The Mandarin Oriental, overlooking the bustling and colourfully lit bay of Hong Kong.

After an amazing roundtrip across the bay on the fascinatingly antiquated and rickety ferry packed to the aromatic gills with a curious selection of local people all struggling to make their way home, loaded up and weighed down with an array of packages and parcels, we returned to the hotel and reconnoitered the plush restaurant that sat at the pinnacle of the building overlooking the night life of the city. Informing the maitre d' that we were on our honeymoon, he assured us of a prestigious table setting in the best position in the room. We returned at eight o'clock to be escorted to our table to find that the promise had been kept. The view was really quite breathtaking; watching all the little toy vessels moving about the inky black water and the neon splendour of the buildings and giant advertising twinkling in the night, all this combined with a culinary experience that has never been surpassed, our evening became a magical memory that remains with us to this day. Returning to our honeymoon suite warmed by a combination of good times, good food, good wine and good cognac, we excitedly anticipated a night of... sensitive, tender, sweet romance.

Hours later I stood on the balcony of the suite surveying my surroundings whilst Annie was still in the bathroom taking a shower. Directly opposite our hotel and right on the edge of the water line stood a row of extremely impressive contemporary office buildings, very much taller than our hotel and majestically stunning against the velvet blackness of the night sky. My eyes were drawn to something in particular; in the space between two buildings way up above my present eye level were two balls of light that seemed to separately bounce up and down with no apparent means of support. My immediate response was to look for a logical and rational explanation for this spectacle. I could see nothing to suggest that this was being artificially created in any way. Excitedly I called out to Annie to join me quickly on the balcony. Rushing out with a towel wrapped round her, my wife looked directly up to where I was pointing.

We watched as the two lights slowly rose upwards, apparently independent of each other, then moved outwards at right angles and dropped rapidly down again. Once more they moved higher, but this time twirling around one another in a spiral that grew in velocity until it reached a point where one of the illuminated balls stopped dead still, whilst the other continued on its path. Finally it slowed to a halt and seemed to just bob as if riding some invisible wave of air. Moving horizontally to the left it once more came to a rest. Rapidly shooting directly up towards the higher light, the lower ball of light appeared on a collision course to impact with the other. The higher light immediately started its descent and seemed to ferociously impact with the other, sending both of them off in opposite directions. Throughout all of these proceedings there was absolutely no sound, especially noticeable at the apparent point of impact. Estimating their size, Annie and I both agreed later that they appeared to be half the size of our clenched fist held out at arms length; I have no idea what size that would actually have made them but I guess, based on the fact that they were, when stationary, parallel to windows that looked at least four foot square, that was more than likely a good estimate for their size. But who knows?! I could be way out!

As we continued to observe the amazing light spectacle, these illuminated orbs constantly rushed up and down, bobbing and weaving, swirling and whirling, bumping and bouncing; it really was quite magnificent to watch and we were adamant that we were seeing something way beyond the norm, be it extraterrestrial or paranormal. We were convinced it was not some sort of man-made light display. Of that we had little doubt.

Finally, one of the lights shot off round the back of one of the buildings, only to be followed moments later by the remaining light, like a cat chasing a rapidly departing mouse. We waited for them to emerge from around the other side, but we waited in vain because they appeared to have totally vanished. We turned to each other in complete incredulity and exhaled with a deep whistle. Raising my eyebrows, I whispered: "Wow, that was some sort of gift. They knew we were here; what do you think?"

Annie agreed. "What a way to commemorate our visit to Hong Kong."

"Our problems are man-made, therefore they may be solved by man. No problem of human destiny is beyond human beings."

John F. Kennedy

Food For Thought: A Vanguard of Change

What has become known as the '2012 Phenomenon' comprises a range of eschatological beliefs* that cataclysmic or transformative events will occur on 21st December or 23rd December, which is said to be the end-date of a 5,125-year-long cycle in the Mayan Long Count calendar. As usual though with any harvest of information it is necessary to segregate the kernels of truth from the deceptive, ambiguous or false chaff. Therefore, I would suggest that if you are interested in the subject there is a wealth of information out there on the Internet, but as always please remember to be circumspect with what you read: *"Just because it is written down and repeated often does not make it the truth"* (Annie Jones – 2010). Always listen with your heart and use your intuitive sense to know what sounds and feels intrinsically right for you. If the 'whole' truth does not reside in any one location you can be sure that 'the pieces to the puzzle' are waiting for you to put them together.

Various astronomical alignments and numerological formulae related to this date have been proposed, but none have been accepted as yet by mainstream academia. A widely acknowledged interpretation of this transition posits that during this time, Earth and its inhabitants may undergo a positive physical or spiritual transformation, and that 2012 may mark the beginning of a new era. Others suggest that the 2012 date marks the end of the world or a similar catastrophe. Scenarios hypothesized for the end of the world include the Earth's collision with a passing planet often referred to as the 12th planet by authors such as Zecharia Sitchin or Nibiru (www. crystalinks.com/nibiru.html) or a black hole in space (http://www. space.com/blackholes), or the arrival of the next solar maximum, which is the period of greatest solar activity in the solar cycle of the sun:

(http://en.wikipedia.org/wiki/Solar_maximum)

Scholars from various disciplines have dismissed the idea that a catastrophe will happen in 2012, stating that predictions of impending doom are found neither in classic Maya accounts nor in contemporary science. Some mainstream Mayan experts state that the idea that the Long Count calendar 'ends' in 2012 misrepresents Maya history, whilst other informed minds would have you believe otherwise. The modern Maya, on the whole, have not attached much significance to the date, and the classical sources on the subject are scarce and contradictory suggesting that there was little if any universal agreement among them about what, if anything, the date might represent. Astronomers and other scientists have rejected the apocalyptic forecasts on the grounds that the anticipated events should be predicted by astronomical observations, which they claim they are not at this time. However, one only has to look at the work of David Wilcock (www.divinecosmos.com) to see new scientific evidence showing that the rest of the planets in our solar system are currently experiencing changes to atmospheric conditions similar to those on Earth. This does not suggest that apocalyptic events will now occur, but it does prove that we are not always being given the complete picture by the 'main stream' scientific community. Therefore, in some circumstances, not having all the 'kernels of truth' might be referred to as dis-information.

So before I go any further, I feel it is my responsibility to open up a sidebar topic that I believe is relevant to the subject of 2012. I don't want to dwell on this point overly long and I certainly do not want our overall subject to be tainted with either unnecessary paranoia or incapacitating conspiracy, but this detour is necessary and will, I assure you, have a positive conclusion.

Based on knowledge that I have received from an unquestionable, irreproachable and totally reliable source, I know there to be human agencies abroad that are aware – to some lesser or greater degree – of the significance of the coming events of 2012 and have it in mind to try and either stymie its positive transformational effect or stop it all together.

Let me tell you right here and now that that will certainly <u>not</u> happen.

Whatever their plan is between here and there, our true human destiny in this matter is not to be tampered with, because there is a higher intelligence at work that will protect the final outcome. When this window of time is finally opened for us, I know that what transpires will indeed be of profound significance to us all. I appreciate that the gravitas of this statement is possibly weakened by not divulging its source, but I stand by my decision to protect its origin when to reveal it would not benefit anybody.

There are two noteworthy and surprisingly necessary components to our life on this planet that we need to be consciously aware of at all times that relate to this subject: Love and fear (not love and hate as some might believe). Our true goal is always to live in Love, because with Love there is truth. But when we live in fear it is possible for outside influences to control our destiny and there are people whose mission it is to manipulate us in that manner. Therefore, to remain in control it is necessary for these individuals to promote fear through lies. One doesn't even have to go to war anymore to make people fearful. One only has to suggest the possibility and isn't that what terrorism has become? Of course, there are still acts of physical violence against other human beings, but they no longer need to be drawn-out affairs, they just need to achieve a certain goal. Whoever instigated the attack on the World Trade Center in 2001 caused a ripple of fear that is still resonating around the world to this day – that was their immediate and long-term goal: to have a devastating and enduring affect upon the human psyche. To them the murder of thousands of innocent souls was merely a means to an end. There might have been side-effect benefits for certain elements associated with this evil cabal – suggestions of what they might have been can be found in a million and one conspiracy theories circulating around the Internet – but the real goal was planting a cancerous seed of fear in the psyches of every single human being. A similar planting is going on at the moment (2010) with the economic climate throughout the world. A very basic fear is being stimulated – our fear of not being able to keep a roof over our heads, clothes on our backs and sufficient food to survive. Because of this financial depression – directly through the loss of income or maybe just by seeing it happening to others – a debilitating fear is strangling our sense of Love. And don't be naive enough to believe that this is a coincidence happening right at this time. It has been building in earnest since 2001, but has been part of a much greater plan for millennia, a plan to control and manipulate the masses in a way more fiendishly corrupt than we could possibly imagine. But the power behind this strategy – which is, by the way, coming from a very simple source: plain common-all-garden depraved and immoral humans (and certainly not from any exotic source) – is beginning to falter, because we are waking up. In fact, we have been waking up for quite a while and although there have been setbacks along the way, we are now stronger and more determined to fulfill our destiny of living with Love than ever before in our history.

Now, here's the thing. The people that would have us living in fear are fearful themselves. Sounds strange doesn't it? If they do not operate from a place of Love, by definition, they are coming from a place of fear. So what are they scared of? Is it Love? Do they believe that Love will incapacitate them; that it makes them weak in some way? The truth is actually quite the opposite in fact, because if you are coming from a place of fear it makes you weak – look at how they are attempting to control us through fear. Therefore they will ultimately fail in their plans, because we can no longer stay rooted in a constant state of dread and impotency generated through their 'smoke and mirrors'.

What the enigma of '2012' represents has never been about the world ending as has been depicted in the disaster movies of recent years. There will not be any earthquakes, tsunamis, comets hitting the Earth, plagues or any other catastrophes of that nature, no more, that is, than in the normal scale of things. It will definitely be a time of change, a time of transition, but it will be something to embrace, not something to be fearful of. It is in the nature of the human race that change does not always sit comfortably with us, especially when it is an unknown change, but life is all about change, for everything about the world we live in is constantly in transition from one state to another. But we have been taught – brainwashed, you might say – to be content with what we have, to never strive to better ourselves, to never explore beyond the safety of our limited parameters, but to be grateful for the crumbs that we are handed. I'm not talking about our material world – for we are constantly encouraged to strive to 'own' more stuff (whatever the cost to ourselves) – I am referring to the spiritual exploration of our humanness. We are constantly steered away from the Love, unless it's under the strict rules and regulations of certain governing bodies, such as an organised belief system for instance. Fear is what is encouraged. It's there every day, right under our noses, being piped into our brains through the media and every other tool of dis-information created to maintain a certain level of fear, so as to control, so as to stop the Love.

We now see that the components of their plan are obvious – they are all fear related. But our arsenal is all Love based and none more important than the influx of our children at this time. It is no coincidence that the schooling for our children has become wanting, that classes have grown greater in size and the overall quality of education has dropped. There is an attempt to dumb our children down. Particular interest is being applied to this new generation, these new special children, the bright and shiny wave of children that are being referred to as Indigo Children, Crystal Children, Star Children, amongst a host of other names.

The term 'Indigo Children' originates with parapsychologist and psychic, Nancy Ann Tappe, who developed the concept in the 1970s. In her book *Understanding Your Life Through Color* published in 1982, Tappe describes the concept of Indigos, stating that during the mid 1960s she began noticing that many children were being born with indigo auras. The 1998 book *The Indigo Children* later popularized the idea along with *The New Kids Have Arrived*, which was written by husband and wife self-help lecturers Lee Carroll and Jan Tober. The promotion of the concept by Carroll and Tober brought greater publicity to the topic; soon their book became the primary source on Indigo Children.

One of the leading experts in the field of Indigo Children is now the British born author and researcher Mary Rodwell. Mary, who now resides in Australia, is the founder and principal of ACERN (Australian Close Encounter Resource Network) and travels the world lecturing on the subject of ET contact and Indigo Children. Through her extensive research she believes that many of the Indigo Children have telepathic abilities and have been spiritually awakened. Mary believes that her research explores evidence from a scientific, biological, psychological, anthropological, spiritual and historical perspective and supports what she believes is a 'genetic' engineering programme for 'upgrading' homo sapiens, leading to a paradigm shift in human consciousness, which the children demonstrate. Her research is now at the leading edge of this phenomenon and includes the latest DNA research, which is helping to qualify how this upgrading may be occurring. Mary also believes that data suggests these children are being altered and transformed on many levels through extraterrestrial encounters.

When 2012 finally arrives are we all going to be given the chance and the ability to make an upgrade that will see the human race fulfilling its true potential? The answer is a resounding yes, though the transition will not be on a physical level, but will come through a vibrational and spiritual shift in our consciousness. The Indigo Children are the vanguard of that change, a shift into a higher consciousness and will continue to illuminate the way for us as we continue on that enlightened evolutionary path.

Part Four

Transformation

"Two roads diverged in a wood,
and I took the one less travelled by,
and that has made all the difference. "

Robert Frost

Experience: The Leeds UFO Conference & BUFORA

Northern England & Central London – 2001

Prior to the year 2001 my memory worked differently from how it works now. That is to say that the ability to access the specific part of my brain or external energetic memory field (the human aura) that contained the conscious and subconscious recorded events of my life was different. I believe that there was a type of security lock on it. You might even have called my memory selective, yet beyond my control to manage. Indeed, this design might be true for all of us, for there is a world of experience that goes on beyond and outside of our waking existence – when we 'dream' for instance, (for want of a better expression). But apparently the memories from these unconscious experiences must remain secret to our waking world, for the fear of them becoming too much of a diversion and overwhelming and influencing our day-to-day life. It appears to me that there has always been a very definite game plan to our life on this planet and especially so at this stage in our evolution. Therefore, being able to 'see behind the scenery' whenever we choose to has never been an option, or so it seems.

There is, and always has been as far as I am concerned, an independent and separate intelligent consciousness at work that has had an input into my/our evolution and life purpose. One might call it a guardian angel; for its 'work' could be seen as a vocation of that nature, as understood by those who subscribe to that belief. But in reality, the consciousness of which I refer to is very real, like you and I are real. It might be referred to as alien, for that is what it is to all intents and purposes, but that is only because we have forgotten our true origin and our connection with that initial point of creation and purpose. I feel that ultimately there has always been a guiding force that has continually offered nurture and counsel within the clear and precise parameters of free will – which is itself a unique and precious gift given to the human race.

There is an array of many and varied definitions when it comes to so-called 'alien life'. Yes, there are beings that originate from physical locations completely separate from the Earth, as well as places that might commonly be referred to as different dimensions – and their location might be right under our very noses. Then there are beings that might be called – based on our lack of understanding and appreciation for the true nature of reality – travellers in time. It is my understanding that we share our space on this planet with many different species; some of which are clearly visible and appreciated within the limited knowledge of what we understand to be 'life' – intelligent or not – and then there are other types of life. We might not even recognize these other variations of life: in fact, they might be going about their business with as little awareness or acknowledgement of us as we have for them. I have a sense that what we are constantly referring to as 'aliens' might in fact be quite the opposite of that viewpoint. Could these 'aliens' actually be long-term inhabitants of this planet, maybe even the earliest and original occupants of the Earth? Can I suggest that there might be several different types of original occupants of what we consider to be our home? We do tend to keep a very tight and constrictive lid on what we consider to be intelligent life, therefore the primary species on this planet might not even be physical beings, but energetic and visible outside of our limited viewing light spectrum. Of course, we need not go far into many cultures to find reference to beings that might fit this description. And then it is worth considering that if a species of human being had pre-dated us and had reached some sort of mental and spiritual plateau far in advance of where we find ourselves now, they might have chosen – for a multitude of possible reasons – to hide themselves away, maybe in a physical location such as the interior of this planet or could have changed their vibrational rate – manually or organically – so as to appear invisible to us.

This appreciation of what we understand 'intelligent life' to be is a constant thread throughout the narrative of this document and remains a key factor if we are ever to fully grow as a species or to begin to see what our 'ET contact' is all about. For is it not true that until we can begin to peacefully live with each other – the many varied and different types of human beings, with our diverse belief systems, cultures and ways of living – how can we possibly consider ever having face-to-face contact with a race of 'alien' beings, let alone experience an ongoing and working relationship?

It is my belief that the subconscious mind protects the conscious mind from what it feels it cannot deal with; one of its tasks is to ensure that we are able to function properly in the day-to-day world without continually re-living an incident that defies explanation. It shields the person from having to accept and process painful or uncomfortable memories or something too challenging, until such time as they are ready to deal with them. I believe the correct term for this is departmentalisation. Everything that happens in one's life is experienced and recorded – if we are aware of it or not – but the ongoing accessibility depends on the criteria set down by this protective subconscious mind. Up until 2001 my protective mind did its job extremely well and I appeared to have no conscious control over it; it worked on automatic pilot as it had been programmed to do so. Accordingly, as events happened I would experience them and then, based on some unknown mental tick list, they would be recorded, processed, then stored away and my ability to access them would then be determined. The one problem with the system is this: although certain memories remain inaccessible, their ability to influence and affect the belief system and physical health of the individual continues unseen. Of course, this whole process went on without any realisation on my part.

It would only be after 2001 that I acquired the 'password' to this particular folder stored in the recesses of my organic/quantum computer. I know now, for sure, that an independent intelligence, outside of my own, deemed it necessary and productive for me to do so. I believe that a physical procedure was carried out, along with other more subtle techniques, to reprogram, amongst other working components of my body, my memory system. Additionally, I was issued with a new conscious stream of information that was to aid my 'dealing' with this reawakened memory facility, along with being able to work within a whole new set of paradigms, goals and instructions. This stream of information continues to this day, where I appear to be downloaded with information in the form of symbols that I can actually see as the process is carried out. The meaning of the symbols is apparently beyond my conscious analysis – although there is a familiarity regarding their make-up – but I believe that another part of my processing system is actually deciphering and assimilating it accordingly.

From 2001 I have been able to have full and unhindered access to an increased section of my memory banks; events and incidents previously half remembered, vaguely recalled or completely forgotten have come rushing back like a movie playing on fast-forward. The images of a thousand events flooded my mind like an emotional tsunami; pushing, shoving and totally rearranging the previously tidy arrangement of my familiar belief systems. Nothing, as they say, would ever be the same again.

In November of 2001 I was made aware of a UFO conference that was going to take place in the north of England, organised by Graham Birdsall, the editor of *UFO Magazine* – the only newsstand magazine of its type at the time. Up until that point, although I had always been interested in the subject matter, I had remained an armchair participant and had never joined any organisations or attended any gatherings. My wife, Annie, insisted that we go, drawing upon an intuitive need and want that would not remain exclusive to this particular event. Without even investigating the line-up of speakers, I booked tickets to the conference and a plush hotel in the centre of Leeds. Upon arrival, I encountered quite a surprise: the main thrust of the event would be Alien Abductions. Now, after reading my story so far, you might find it unbelievable to discover that I was not interested, at all, in the subject matter of Alien Abductions! In fact, I used to go out of my way to avoid the subject entirely. When reading a magazine or a book on UFOs where abductions would be touched upon, I would simply flick past the relevant area and quickly move on. I now clearly see that at that time a part of my subconscious brain was actively protecting me from having to encounter and deal with anything to do with this type of event; a subject matter that had the potential to bring about pain and anguish, based on the distorted interpretation of an event that had been coloured by my limited, misguided or retarded belief systems. As I have moved through life, my reaction and response to my ET contact has fluctuated; reflective of the age, emotional, spiritual and intellectual level I had reached at that particular moment.

After a series of varied events that had been taking place throughout the year, there were long-forgotten treasures rising to the surface of my emotional sea, with many new and exciting aspects of my psyche beginning to gently nudge and filter into my conscious world. Therefore, it was with a sense of initiation that I arrived at the UFO conference. Additionally, this sense was supported by our joint awareness, although yet unspoken, of Annie's clear urgency for us both to attend without identifiable reason beyond the obvious. I came to acknowledge and work with her intuitive sense, but at that time I did experience a degree of reluctance based on, I suppose now, my need to be safe within the confines of my 'normal world' and not to push the envelope of that experience.

I want to make something very clear: what happened at the Leeds UFO conference was a catalyst; a tool that was used by an unknown force to prompt me into taking steps along a new path. My contact with a specific member of the presentation team was important, but I do not believe that we were connected on any other level than her purpose in lighting a touchpaper that ignited the beginning of my new life-to-be. With that said, I now feel more comfortable in the telling of events over that weekend.

So it was with mixed emotions that Annie and I filed in to the conference hall and took our seats on Friday night for the opening introductions. Graham Birdsall, founder and editor-in-chief of *UFO Magazine* and compere for the events introduced Nick Pope. Nick had been working for the UK government, allegedly sifting through, analysing and reporting on all UFO sightings made in the UK. Affectionately, he had been labelled the British 'Fox Mulder' after the fictional character on the long running television programme *The X Files*. He spoke briefly about the talk he would give the next day in conjunction with an Experiencer called Bridget Grant. In line with my lack of knowledge in this area, I assumed that an Experiencer was in fact an Alien Abductee, and I was, in fact, correct because all that had happened was an updating of the labelling system. As Bridget joined Nick Pope on stage, I immediately thought I recognised her. This came as no real surprise as I believed I must have seen her picture in a book or magazine on UFOs. That was until Nick announced that Bridget was making her first public appearance and had never received any media coverage before. I stared at Bridget, rationalising that she must just look like somebody I knew or had seen before. Before I could really think about the matter any further Nick and Bridget were leaving the stage with a promise of what was to come in their full presentation the next day.

As I lay in my hotel bed that night finally drifting off to sleep after an intense day of travel, the last thought I had before I finally succumbed to sleep was that Bridget wasn't Bridget; I definitely knew her by another name. And then I was asleep.

Refreshed and ready for the new day, Annie and I arrived at the conference centre and took our seats. First up would be Nick Pope and Bridget Grant. The events of the previous day were still nagging at my logical mind, but I had pushed them far enough back so as not to be distracting. That was until Bridget once again took to the stage and began to tell her story.

My focus was totally on the young, dark-haired woman who was speaking. I no longer felt that I was in the auditorium; everything around me seemed to fade away. Just as if my eyes had the ability to zoom in on what I was looking at, Bridget's face moved closer into focus. As she outlined specific experience details concerning her alien contact, with each sentence that she spoke I somehow knew what she was going to say before she said it. My mind pre-empted the sequence of events as they were unfolded. It was as if I had heard them all before. My mind screamed at me for explanation as each second passed. What I was experiencing became more and more unbelievable. All I could see was her face; all I could hear was her voice. Whilst she was still relaying the closing moments of her experience, my memory of what I had somehow heard before had already come to its completion. Bridget smiled, placed the microphone in her lap and looked down at the floor as the audience began to show its appreciation.

The hypnotic trance that I had found myself in was broken by the crackle of applause and I suddenly became aware of my surroundings once more. Annie was tugging at my sleeve and was asking if I was all right. Looking at her as if bewildered by the question, I smiled in response and robotically began to join in the applause. An announcement was made to signal a short refreshment break in the proceedings and people began to filter from their seats into the foyer. I did not move. I just sat there gazing down at the empty stage. Once again, Annie asked why I appeared to be acting so strangely, but I responded with little regard to what was being asked of me and just stared ahead blankly, nodding vaguely in reply. It felt as if a million questions were being fired in my own mind, all at the same time. As one question was parried, another was waiting in line to be explained. But there was no explanation. What I had just experienced was impossible. I knew that. My logical conscious mind knew it was impossible.

Without me realising it, Annie had left her seat and gone off to purchase refreshments for us both. Upon her return she found me still staring blankly ahead. Handing me a bottle of mineral water she playfully waved her hand in front of my face and enquired if there was 'anybody home'. In that moment some sort of survival instinct kicked in. I would not be totally, consciously aware of what happened, but my eyes began to focus and I was 'back in the moment'. Somewhere in the deep recesses of my brain some sort of in-built safety switch had been thrown. As the impending overload that was the reality of my experience began to threaten my ability to manage the situation, the brain's aptitude to departmentalise what it cannot cope with took over. On one level, I realised that I believed I had recognised Bridget Grant and that I had somehow known what she was going to say before she said it. I realised it, but I did not understand it. I knew that I had experienced it but I did not believe it had actually happened. Under those confusing conditions the brain did what it had to do to protect me.

I took the bottle of water from Annie and asked her who the next presenter was. For the rest of the day and that evening, I acted as if nothing was wrong. The experience had been squashed away in a place I wasn't able to reach. Therefore, on a purely conscious level it didn't really bother me. However, as with all things that we try to suppress, it continued to nag away at me on a subliminal level. There were moments when my mind began to drift and I was unable to concentrate on what was being said to me. Annie put it down to tiredness after the long drive and the intense concentration required for each presentation. There was a part of me that knew differently though. And that part was struggling to get back out.

Over the next two days I kept noticing Bridget Grant: in between lectures, moving about the lecture hall, in the reception area, queuing for refreshments – everywhere I seemed to go I saw her. She seemed to scurry from one location to another, always with a young man in tow one or two steps behind her. I never caught her eye, in fact it appeared that she saw nobody else, as if unaware of her surroundings, but always with a determination to reach somewhere other than where she was. As I stood in the main entrance Annie came out of the toilets and told me that she had just seen 'that abductee girl' washing her hands; she went on to comment about her apparently 'nervy nature' around other people.

Throughout this time, unexplainable memories seemed to be pushing their way back to the surface of my conscious mind. Somehow or other, when I wasn't occupied with the distraction of each lecture, I wrestled with the increasing awareness that I appeared to have knowledge of Bridget Grant without apparently ever having met her before. As each detail presented itself, my rational mind instantly rejected it. Conflicting inner voices fought to be heard, whilst all the time my outward appearance showed no real sign of what was raging inside of me.

On the final day of the conference it was announced that all of the speakers would take part in a question and answer session at the end of the day. I immediately announced to Annie that I would be posing a question directly to Bridget Grant if the opportunity arose.

"What are you going to say?" my wife enquired.

"I'm... not sure yet," I hesitated. But I did feel a sense of relief at the prospect of doing so.

As the day progressed, it was very clear that there was not going to be any time to fit in the additional questions and answers element. In fact, due to the organisers' handling of events, we would be lucky to hear all the speakers before the end of the day. So I knew that I would have to find an alternative window of opportunity to speak with Bridget Grant directly. I mentioned this to Annie and she appeared to intuitively appreciate the importance of what I was saying. She encouraged me to speak to her in the final break of the afternoon before the last speaker of the day. Although I still wasn't sure what I would say, I knew that I wouldn't be able to leave that day without doing so.

Just before 3.30 p.m., my concerns were proved to be accurate: it was announced that there would be no question and answer session. In fact, the final break of the afternoon would have to be shortened to accommodate completion of the day's programme. My eyes scanned the front row of the auditorium that seated the conference speakers. I immediately saw where Bridget Grant was sitting and started to make my way down the steps towards her.

As I approached I saw that she was already speaking with somebody. Reluctantly I stood to one side waiting for her to finish and hoping there would still be enough time to have a few words myself. As if sensing my presence she turned her head to the right and looked straight up at me. At that instant, as our eyes made contact, I saw what I knew to be recognition reflected back. In that moment Bridget Grant's eyes said, "I know you". As quickly as it appeared though, in the next instant it was gone. Although the smile that had accompanied the recognition remained, there was now a look of hesitation. Her brow furrowed as she continued to look up at me.

"Hello," I said. "Would it be possible to speak with you?"

"Of course."

She didn't look back at who she had been talking to, but continued to return my smile. The person sitting beside her slowly got up and relinquished their seat with obvious embarrassment and awkwardness. By the time I sat down, although Bridget Grant's face remained friendly and welcoming, I don't believe that I saw any recognition remaining in her eyes at all. Whatever plan I had with regard to exactly what I was going to say left also. We spoke for about ten minutes and during that time I used certain terminology and phrases that would suggest I had some prior knowledge of her character. Each time I did so, I waited for Bridget to question my particular use of words and grammar, but she failed to pick up on these points. In fact, she appeared immediately at ease with me and we spoke with a comfortableness of old friends. I thanked her for her brave contribution to the weekend and bade her good luck and farewell. By this time our meeting had generated a level of emotion for Bridget that found her tearful (for reasons only known to herself). She then took my hand in hers and stated, "Something has happened to you hasn't it?" I still remember the moment with crystal clarity because I heard a voice say, "Yes it has". It was only as I saw the tears begin to trickle from her eyes in response did I realise that the reply had come unknowingly from my own lips. My final comment was something like, "Be careful," or "behave yourself" and her reply as I turned away was, "I always do, you know that." I sat back down in my seat stunned; what had just happened? I looked back down towards where Bridget was sitting to find her looking back in my direction. The strange thing was, although she sat in the front row and I was way back in the upper rows it was as if her face raced towards me and I could see her eyes clearly as they met mine.

Addendum

What happened to 'Bridget Grant' you may ask? Three people I spoke with a short time later confirmed that was not her real name, but were unable to furnish me with any further information that would enable me to contact her. Nobody appeared to really know much about her apart from what she had presented through her 'contact experiences'. Budd Hopkins did look into finding her for me based on the last known address he had for her, but his search came to nothing. In his words, and I don't mean to make them sound mysterious in any way, he said that she had just apparently *disappeared.*

My feelings are that 'Bridget' had had her moment: that she had taken the opportunity to speak publically about what had happened, for whatever personal reasons she had, then went back to her 'real' life. It's a shame I never had the chance to speak with her again. I don't believe that our meeting at the conference was anything more than one of many catalysts that occurred for me to begin my journey of self-discovery, but I did sense there was more to be said to each other. I also sensed that the ETs had played a part in bringing us together, in stimulating my peculiar response when hearing her experiences and to prompting my change of direction as far as an *interest* in so-called alien abduction.

But as for the million-dollar question: Did I believe that 'Bridget Grant' was 'Little Star?'

Budd Hopkins did. After we first worked together in New York he said that he would try and track Bridget down because he had an inkling that she might turn out to be the young girl in my contact experiences. At that stage I still felt it might be a possibility; for although I hadn't seen Little Star for quite a while I imagined that she would now appear very similar to how Bridget Grant actually looked: roughly the same age group, slim and petite, long dark hair. But it would only be a matter of months before I would be able to confirm *beyond any doubt* that Bridget Grant was *definitely not* my Little Star.

BUFORA

I don't think it would be an exaggeration to say that what happened at the Leeds UFO Conference knocked me for six. Over the next few months I began earnestly to read-up on the subject matter of alien abduction through the work of Budd Hopkins and Dr. John Mack. Initially after investigating the UK UFO scene, I quickly realised that unlike in the USA there was absolutely no support system in the UK for people who had experienced contact. Although Annie remained firmly by my side throughout this period, my journey of discovery and acceptance was a solitary one. We attended a one-day UFO conference in the suburbs of London conducted by the British UFO Research Association (BUFORA) and, although it was an enjoyable day out, on the whole we were somewhat disappointed at the content of the event. Although I only remember one decent presenter, my heart really began to sink when a well-known debunker took to the stage, prattling on regarding his negative and destructive viewpoints.

I could never understand why he was there in the first place.

During the day I spoke briefly with the chairperson of BUFORA, Judith Jaafar, concerning an investigation that was going to be carried out on abductions. I agreed that I might be interested in participating and she said that she would be in contact. It didn't take me long to discover that the investigation was going to be conducted by somebody called Dr. Chris French from the Anomalistic Psychology Research Unit, a well-known skeptic. Now, I have nothing against skeptics as long as they are caring, intelligent and constructive people and as long as they acknowledge both sides of the story, but Dr. French, I found out, had a reputation for attempting to make abductees look stupid in a very clever way: he does this by allegedly selecting only data that fits his debunking parameters and ignores the rest. Another dead-end!

I advised Judith of my intention to withdraw from any investigation but she wanted to know if I would be interested in conducting a presentation about my experiences in London at a small gathering of BUFORA members. After a short deliberation I agreed. A couple of weeks before the get-together, Judith invited me to her home one Saturday afternoon to talk about my experiences with just her and her partner. Whilst I was there a friend of Judith's, Malcolm Robinson, a Scottish UFO/Abduction investigator, author and presenter, paid a brief visit. We spoke for a few minutes unaware at the time that our paths would cross in the future, where I was to discover that Malcolm is one of the hardest working investigators in the field of UFOs and Metaphysics, and is one of the most entertaining presenters I have ever had the pleasure to listen to. In recent months (2010) I have also become aware that Malcolm is about to launch the first newsstand UFO magazine (*UFO Matrix*) in the UK since the demise of Graham Birdsall's *UFO Magazine* nearly 10 years ago, along with Philip Mantle, another author and investigator and I wish them both well on their new venture: UFO Matrix (http://www. healingsofatlantis.com)

During my afternoon talking with Judith and her partner, I was to discover that she was an extremely healthy skeptic (although I wonder if she would agree for she did keep it very well camouflaged). Our supposed comfy chat was more of a friendly *interrogation*, but I was extremely comfortable with the challenging format and understood where Judith was coming from: she had to make sure I was the real deal. A little while after this Judith was to confide in me that after I had left she telephoned a colleague to tell them that she had just interviewed her first genuine and authentic alien abductee.

As for my London presentation, it was very much a baptism of fire in a lot of ways. A great many UFO enthusiasts are very much like traditional trainspotters (sitting on train stations, just noting numbers), in that they are very happy to observe unexplainable lights in the sky and talk about them endlessly. However, what is so peculiar is that they have no real desire to discuss the possibility of there being occupants within the crafts or to go further and investigate the purpose these beings have in their contact with us. And god forbid if anybody should propose what that contact means to the human race and how we should respond. Let me say it loud and say it proud: the contact that we are all experiencing is happening for a reason!

"What we can or cannot do,
what we consider possible or impossible,
is rarely a function of our true capability.
It is more likely a function of our beliefs about who we are."

Tony Robbins

Experience: Locked Out of the House

Essex, UK – February 2003

On the evening that this experience transpired, Annie and I had retired to different bedrooms because of my ongoing insomnia and the need not to disturb my wife while I read before finally falling to sleep. Having to do this was not a regular occurrence, but I was experiencing a lengthy bout of interrupted sleep, so I took the opportunity to sleep in my son's bedroom, which was vacant at the time. Annie was never happy with my vacating the matrimonial bed but I was very aware of my reading light being on and my restlessness whilst she tried to sleep. So off I crept....

I remember lying on top of Reece's bed because it was my intention to return to my own bed when I felt drowsy enough to successfully fall asleep. Although the night was bitterly cold outside, the house still remained cosy and warm; the remnants of the heat generated from the central heating system being on all day. As such, I wore only a t-shirt and a pair of underpants as I lay there reading my book.

After a while I was distracted by an irregular sound I heard emanating from downstairs. Believing I had no other option but to investigate I put on my son's dark blue towelling dressing gown and stood at the top of the stairs listening intently. I felt reluctant to investigate further, even though my mind kept presenting me with unavoidable reasons to explore the origin of this unidentifiable noise. Still I hovered, unwilling to proceed. Below me our dog lay motionless stretched out on the wooden floor of the downstairs hallway, illuminated by the streetlight through the glass of the front door. This in itself was peculiar as she would normally have stirred by now from the movement of somebody walking about upstairs, but she appeared to be undisturbed and apparently slept on regardless. The idea that the sound coming from downstairs might suggest something ominous did not finally provoke my investigation, instead it was an overwhelming need to quench a newly realized thirst that moved me ultimately to explore beyond where I stood. Reluctantly, I descended the flight of stairs whilst all the time trying to find sufficient cause to return to my book. I never once questioned this absurd reasoning or that my normal reaction would have been to immediately identify the source of the initial disturbance. Stepping over the slumbering dog at the bottom of the stairs, I half-heartedly pushed open the lounge door and stepped inside. Foolishly, and for some unknown reason, I did not put on the overhead light. I scanned the darkness hoping to see an understandable cause to explain the disruption. All at once a movement caught my eye. Looking towards the sliding glass doors at the end of the dining room I could see something was moving on the paved patio area right outside the window. In an attempt to distinguish exactly what I was seeing from the inanimate objects in the garden, my mind worked hard to process the possibilities. In what seemed minutes, but probably took just a moment, I had exhausted all probabilities. The movement had come from two or three external sources; they were upright, I would estimate at three or four feet in height, slight in size, and definitely animated, albeit moving jerkily. My conclusion in that moment – although now really quite ridiculous – was that I was witnessing foxes walking upright on their hind legs!

I took a couple of hesitant steps forward. As I did so the 'foxes' turned towards me and instantly froze. There were three of them; they had sharp, bright eyes that seemed to shine in the darkness. I was fixed in their gaze, unable to move any further. I swallowed hard and could feel my jaw clenching. I wanted to look away, but I couldn't. I wanted to turn away and go straight back upstairs to the safety of my bed and Annie, but I just couldn't move. I just stood there, staring at the foxes as they stared back at me. Then something peculiar happened.

As if a thin, gossamer veil lifted I suddenly became aware of what was actually there. The foxes somehow blurred and then re-focused before me; what I had thought of as familiar became something completely different and alien in appearance. What I had initially seen as rough, dark-red fur became smoother in texture and lighter in tone; the 'snout' drew back and became flattened; the 'ears' that had previously extended from the top of their head were simply not there; short, stubby, jointed front 'legs' became long, apparently jointless arms with uncommonly thin pointed fingers, which hung still and lifeless by the side of their bodies; spines straightened from slightly stooped postures; pointed and elongated features became rounder and bulbous. At once I was magnetised by the transformation from 'canine' eyes to something disturbingly unworldly: two overly large, intensely black, teardrop- shaped eyes, wrapped incongruously from the centre to the side of the face.

What I imagined to be the patio floodlight flickered in intensity above their heads and illuminated the scene in a cold, blue-tinted blaze. They continued to stare. As I grudgingly stood frozen staring back the 'thing' in the centre stepped forward. It stepped forward right through the solid plate glass door that had stood reassuringly between them and me. There was no breaking of glass, no sound of cracking; one moment it was outside, then the next moment it was inside. I then watched as the remaining two beings also stepped through the window and stood either side of the first one. I can remember a voice in my head screaming at me in total disbelief to "Run, run, run!" but I remained paralyzed on the spot. One moment they were about 12 feet away from me, then, as if in the blink of an eye, one being stood on each side of me firmly grasping my elbows. The remaining one was directly in front of me; its head titled up towards mine with its abnormally large black eyes seeming to somehow pulse in intensity. I knew it wasn't so, but it felt as if its eyes were completely in-line with mine, merely inches away. It turned and began to walk back towards the window. I felt the pressure on my elbows increase and I was guided down the length of the room, just a couple of steps behind the path of the leading being. Although I didn't appear to have the ability to speak out loud, my mind continued to verbalise my feelings as if I was actually talking: recalling a similar abduction incident many years before, I pleaded passionately for the window to be opened. Memories of what the process of transition through a solid object felt like filled me with uneasy trepidation. I closed my eyes and experienced the familiar feeling of stickiness across my face and the pulling sensation in the area of my solar plexus as I apparently moved through the solid plate glass window. Instantly I realized I was now outside as I felt the freezing air of that winter night upon my face, hands and feet. Through a jumble of mixed emotions and feelings I was aware that my feet were no longer on the ground and that I appeared to be rising up into the air. Reluctantly I squinted my closed eyes open just enough to see the garden disappearing below me. What I had thought to be the patio light was in fact a powerfully unnatural light source beaming down from above me and I was rapidly rushing towards it. The lifting pressure remained at each elbow, so I could only imagine that my accompanying abductors were still attached either side. The sense of familiarity increased, as did my apprehension of what was possibly to come. The feeling of nausea began to grow in the pit of my stomach; whether from my bizarre transit or the anxiety of circumstances to come I did not know.

With my eyes now firmly clenched shut, I was unsure of what then transpired. I experienced movement around me, a change in air pressure that popped my ears and finally a force beneath my feet that made feel that I was standing once more on solid ground. Although feeling slightly wobbly as I stood there, apparently no longer supported by my abductors, I eventually opened my eyes. Wherever I was, plainly was not bright enough to see properly. The 'room' felt cramped with the outline of three small people silhouetted before me; there was a diffused hazy quality to the light that constantly changed my perspective. To the right of me appeared to be some sort of table, which most peculiarly, I then saw in my mind's eye myself lying upon. I realised that 'they' wanted me to lie down and this was how they were telling me. Was this some sort of telepathy? I resisted. I didn't want to lie down because in my experience it was not always a pleasant occurrence. Instantaneously I could feel something holding me, every single molecule of my body was being held and in that moment I began to move. As if a million tiny invisible hands were lifting me I unwillingly moved towards the table, turned and sat down. I succumbed to the inevitability of the situation and consciously relaxed. As I did, I felt their hold relax too. I said in my mind: "I just get scared, that's all. You don't have to be so forceful, you just have to ask and I'll do it. I just get scared. I *know* that it's important". Somehow, and I really don't know how to describe it, I 'felt' their response and it was one of gratitude.

As I lay down on the table I could feel movement beneath me. It was as if the table was molding to the contours of my shape. There was just the slightest of 'give' as my weight pressed down and the surface rearranged itself to best support the curvature of my body. Two of the beings appeared, one on each side of me. One of them reached across and lightly brushed my forehead and in doing so I felt the need to close my eyes once more. I could hear them moving beside me. Then a hand touched my face again and I felt pressure on my nose. There was a rush of images and feelings that swamped my mind, totally distracting me from the moment. Whatever this information was, it was new to my consciousness; all of it, without exception, was information I did not recognise nor appreciate its value. It wasn't a foreign language as such, but it was exotic to my understanding. As the flow of images finally ceased, as if on a movie screen, I saw myself putting something into a box, standing upon a chair and placing that box onto a high shelf. This image made me feel relaxed again. I continued to hear movement about me as I lay unperturbed stretched out on the table. I found it quite difficult to focus on what was actually transpiring, and my mind began to wander to more mundane thoughts. Was I falling asleep? Drifting into a different state of consciousness? Was this train of thought even under my own control? I floated from thought to thought, my mind effortlessly moving...

With a jolt my eyes opened wide. My senses felt immediately overwhelmed: a blinding lightsource; a blanket of coldness; a change in air pressure; and a breath-stopping surge of palpitations. I was standing upright. My head was turned upwards looking directly into an extremely bright light. It seemed so close that I could reach out and touch it. As it began to pull back I could see that the light beam was attached to the underside of an extraordinarily large, round object that was floating in the sky above me. As the light got smaller the object seemed to get bigger. I stared intently attempting to identify what I was witnessing. It was a... UFO?! Slowly, but very definitely, it moved up and further away from me. It was an alien spacecraft?! As it moved away it became smaller and smaller. It's where I had just been?! Up, up into the night sky. It was a... comet?! The light was diminishing. It was a... shooting star... it was just a... star. It was gone. It was nothing.

As if mesmerised, I stood stock still gazing up into the night sky, watching individual stars twinkling in various degrees of brightness. I blinked and as I did I became aware of where I was: I was in the garden behind my house, I was wearing my son's dark blue towelling dressing gown with now nothing on underneath, my feet were bare and cold... and it was night-time. What was I doing here? How did I get here? Had I been sleep walking? I tried the handle of the patio doors but it was locked from the inside. The freezing cold temperature seemed to aid my quick thinking and evaluation of the situation: I had obviously been sleepwalking, I was now locked outside the house and I had to wake my wife up to get back in. All of the windows were closed, so I walked round to the front of the house and looked up at our bedroom window; there was no sign of life, the blind drawn and no light on. Where I stood was a small front garden plot with flowers and bushes. I rummaged underneath the foliage to find a small selection of tiny pebbles that could be thrown at the window to hopefully gain my wife's attention. Before I did this I tried ringing the doorbell, although I didn't expect this to be of much use as I knew it didn't work properly. This turned out to be correct, so there I was, god-knows what time it was, naked apart from my son's dressing gown, lobbing little rocks at my bedroom window! I really hoped that nobody was looking at me out of their window. After a few well-aimed projectiles my wife's sleepy-looking and extremely puzzled face appeared at the window and attempted to focus on the peculiar apparition that stood beneath her. Within seconds I saw recognition in her eyes and although clearly confused she gestured to me that she would come down immediately. I stood waiting, shivering, half-naked by the front door waiting to be able to get back inside. The hallway light came on and I watched as her fuzzy shape could be seen through the glass panel in the door descending the staircase. I then heard a sequence of noises that within seconds began a terrifying response in me. Annie began to unlock the door from the inside. First of all I heard the security chain slip from its holster, then the first lock was turned and then finally the remaining one. What could this mean? How on earth could I have left the house if both the front and back door were locked from the inside? By the time Annie had opened the door the crippling truth of this fact – *its impossibility* – had overcome my emotions. I stammered, "How did I get outside? The doors are locked. How the fuck did I get outside?"

We were both clearly disturbed with what was being presented and expressed that emotion accordingly. Hugging each other in desperation, we stumbled into the hallway and shut the door.

"Check all the windows," I said desperately clutching at straws. Within a couple of minutes Annie was back at my side having found that all of the windows and doors were securely fastened and locked from the inside. It was the middle of a freezing cold winter after all!

Then something most peculiar happened. As if a slow warm wave of calmness passed over me I suggested that everything was all right and we should now just go up to bed. Annie stood there, blinking at me in total disbelief. I took her hand affectionately and, stepping over the dog that was still – *still* – completely motionless at the bottom of the stairs, apparently fast asleep and down for the count, I gently guided Annie up the stairs to our bedroom. All the while I was attempting to calm and reassure her that there was no need to worry about anything. Obviously she just looked back with incredulity, as it was apparently clear to her that she had every reason to worry about what had unfolded. I forcefully, but tenderly, guided her into bed and slipped in beside her, all the time reassuring her that I was OK and that there was no reason to be concerned. I moved over to her side of the bed and wrapped my arms around her quivering body. Eventually, but after quite a long time, Annie's sobs and little emotional quivers began to subside until finally she lay still in my arms. A short time later I believe we both fell to sleep.

When we woke in the morning, we just lay there staring up

at the ceiling. I can't recall what I said, but I did speak about the previous night's events and what they meant. In fact, I know I spoke for about an hour without stopping. When Annie and I discuss that morning-after now, even she can't remember what I said to her. However, whatever I said made us both feel better about the whole situation. I don't think I spoke about what had actually transpired, how I got out of the house and where I went because it would be a while before those memories resurfaced and came back into focus for me, but I spoke about something of a much greater proportion, on a much grander scale, if that makes any sense at all. My sense is that it wasn't really me speaking on that morning, not exactly, not the *me* that operates on a daily basis here and now. I sense that my words came from a different version of me, one that operates from a higher consciousness, and one that has access and understanding of everything that I experience on all different levels.

*"A sufficient measure of civilization
is the influence of good women."*

Ralph Waldo Emerson

Food For Thought: The Experience of the Experiencer's Wife

Written by Anne Ashley-Jones (Annie – my wife)

What is our experience, we, the people who live on the periphery of the human beings who have this contact experience? What does it mean for us living out our lives in the strange and, at times, wonderful arena of the world of the ET Experiencer?

For me, it was not something that I asked for, although I will say that everything I have ever believed in or was interested in from a very early age encompassed so much to do with the concept of extraterrestrial life.

Initially, as Steve's story unfolded and I came to accept that he was indeed having genuine and frightening experiences with no logical or rational explanation – except for the very obvious fact of what they actually were – I was extremely fearful. Not for myself, but for my husband who apparently had no control over what was happening within the confines of his own home or in the world around him. More and more I came to dread leaving the house for any length of time as I was never sure what I would find upon my return. My major fear was that one day he would be taken not to be returned.

This fear meant that I would get up in the middle of the night and, on finding his side of the bed empty, go in search for him to make sure that he was all right; for I could not sleep until I had found he was safe and sound. More often than not he would be tucked up on the sofa fast asleep with the TV on. On returning home from time spent with friends or family I would reach the last turning before our house and instinctively know when something had happened to Steve. Even at that distance the energy would reach me and with dread and trepidation I would step into the house and immediately go to find him upstairs in the bedroom. After one such occurrence I was to discover him with blood over his face and unable to be roused; this was Steve, the lightest sleeper in the world who, if you accidently touched him in the night, would leap up like a scalded cat.

Using a UV Light to show up trace elements left on his skin after an ET contact I began to document the events by taking photographs of all the weird marks on his body – at times distinct fingerprints – and writing down what he was able to remember. I never had any memory of events myself. Even when Steve told me that an ET had led me into the bedroom he was in when we were staying at friend's home, it meant nothing to me. Not until we returned home that is and looked at my arm under the UV Light and saw what appeared to be three long straight finger marks on the arm he said was held by the ET. This was a huge shock to me as it meant that things were being done to me as well as him. I too am an incredibly light sleeper, so it was hard enough to accept that I wasn't waking up to help him, but to think that I was being manipulated as well was very disturbing.

One evening when Steve and I were getting ready to sleep whilst staying over with a friend who had two very young children, he *casually* announced that "They are coming for me tonight," and then promptly fell asleep! Dumbfounded by his statement, I was determined to stay awake; but this was more due to for my fear for the children's safety than anything else. I successfully stayed awake for quite a while, but of course ultimately fell asleep only to be shown by Steve the next morning a very neatly shaven circle on his thigh, which was definitely not there the night before. What was the point of that? To show us that no matter where we went or who was in the house, if they wanted to come and get him they would... and could? I suppose it worked as I accepted that wherever we were – at home, away from home, even in a different country – it would make no difference at all if the ETs were coming to collect him. The fact that so often they came when I was out for the day only confirmed to me that they were in tune with all that we did on a daily basis. I liked to believe that I might have been problematic to them when they tried to 'put me out', so that they could take Steve. I felt this was why they sometimes chose times when I wasn't at home, but somehow I doubt that.

The greatest difficulty for someone in my position, and for Steve himself, is dealing with the disbelief of those people whom we are brave enough to confide in. I understand that if the experience is not happening to them they cannot accept that such events are possible. However, knowing us to be totally honest, reserved, extremely private and intelligent people, I always felt that there should be the question of "Why on earth would they make something like this up?"

Just recently my closest friend told me that she had finally told her partner about Steve's experiences, and his reply astounded me. He said that although he felt that would not be unusual for Steve to believe that these things could happen to him (whatever that was supposed to mean!), he couldn't understand how I could accept it because *"Annie is such a highly intelligent woman, whose knowledge of so many things astounds me, how could she believe such nonsense! That just doesn't make sense to me"*.

Does this then imply that the general belief is that only stupid people can believe in these phenomena or can accept these challenging new dimensions we find ourselves being confronted with? (Tell that to the Dr John Mack's of this world!) The thing that struck me most about his statement was that he felt he could have such a strong opinion on the subject without knowing anything about it at all – a common problem we find. He knew nothing about ET contact or even what our experiences were, yet felt able to say we must be stupid if we believed in what had supposedly happened to us. Interestingly these people feel no need to actually ask you what these experiences are.

So, this is what we find ourselves up against: people who judge us and our experiences and then decide that we are deluded or just plain stupid, their opinions based on no knowledge or information about either the subject or ourselves. Rarely do you find someone who is not involved in the field who will ask you, "Well, I don't know anything about any of this, so please enlighten me and then I can make own mind up".

So we become isolated and scared of who we talk to. We suffer anguish over having loved ones ask us never to speak about our experiences in company in case it reflects on them and people might think they have mad relatives! There is often the experience of some people appearing to agree with us just to clearly shut us up, knowing, that in truth, the person does not believe us at all.

Although all of our friends in America are 100% supportive, the very few friends and family in the UK who show support can be counted on one hand. Having said that, these are the individuals that help me to cope with the negative reaction from the others. Our discussions with friends in a similar situation around the world shows the same pattern and many Experiencers become completely isolated from their families and friends who want nothing to do with them if they continue to state that their contact experience is real and not make-believe. A very sobering and saddening fact.

I, however, continued to bang my head on the brick walls of their disbelief and related new events as they occurred. I will never deny the truth of our experience, apologise for it or try to justify it, as I have no idea, in truth, why the ETs are doing what they are doing or why they chose Steve. I continue to research, to talk to people who understand what we are going through and try to find some answers for the brave and courageous people who are living with this on a daily basis – like the relatives of alcoholics, I feel that the fallout for the Experiencer's family is vastly underrated. The best people can offer us is an acceptance of our reality – without judgment, without hostility and without ignorant remarks.

My only hope is that one day this phenomenon will be totally out in the open; with ETs and their craft appearing on the physical plane as a normal occurrence, and then people like my brave husband will be vindicated and obviously become very sought after, as they will be the ones with the answers and not the elected politicians. I find it interesting how ordinary the Experiencers are in themselves; as there are no rocket scientists or politicians or religious leaders (not that we know of anyway!) When people ask me why don't the ETs just land on the White House Lawn or in front of Buckingham Palace, my response is that they seem to have no interest whatsoever in communicating with anyone who is telling the rest of us what to do.

They only want to contact the normal man or woman and have an interaction with them.

I find this hugely comforting.

"I love you not only for what you are,
but for what I am when I am with you."

Elizabeth Barrett Browning

When it comes to my transformation, Annie has been the pivotal part of the
mechanism that all connected cogs and wheels turned around.
That is why I have devoted a specific chapter to her.

An Introduction to My Annie

There's absolutely no point in attempting to hide this secret anymore because it will only come out eventually. Whatever subterfuge I try to employ to disguise it I know that it is only a matter of time before everybody finds out the truth. I truly abscribe to the belief that if this dark clandestine fact is suppressed any further it will find its own way to break free. So I just need to be brave, bite the bullet and say it out loud. And then it will be out there. No more sleepless nights; no longer will I toss and turn in the wee small hours worried what people will think when they discover the truth; and then, *and only then,* will I finally be released from the torment of harboring this skeleton in the cupboard.

Annie, my wife, my soul partner, my journey's companion, the love of my life, my majestic consort, *does not like desserts.* There I've said it.

She will partake of the occasional crème brulee when the fancy takes her. She will even consume the odd Eccles cake or custard tart when pushed. But that's it. It's not that she doesn't have a sweet tooth, because she has been known to eat the odd square of chocolate and to get her fingers powdery-sweet from the far-eastern promise of Turkish Delight. But she's just not a dessert person. And that's sad because I am.

Reluctantly I have to say that it is the one thing that stops this remarkable human being from reaching pure perfection. It is the singular blemish in her character that hampers the dovetailing of our two souls in eternal bliss.

Let it be known that my darling Annie is a culinary artiste of the highest order when it comes to the creation of all things savoury, but when the need calls for the construction of spectacular sweet pastries and exquisite puddings, her gastronomic skills let her down tragically. We have sought help in this matter and many hours have been spent with her lying on a psychiatrist's couch attempting to remedy the situation, but she normally leaves the frustrated therapist with his head in his hands.

It is a tragic tale, but one we are attempting to work with.

Baby steps are being taken in the kitchen, as Annie opens a tin of creamed rice pudding and deposits the contents delicately in a bowl; the tin foil lid of a pre-prepared trifle is peeled back with wonderful dexterity and presented elegantly before me; an exotic and provocative banana is seductively sliced and covered with smooth Devonshire cream; whilst all the time we are hoping and praying that she will be one day released from this culinary cul-de-sac.

Apart from this inexcusable imperfection in her makeup, we have learned to somehow cobble a happy life together (he-he!). In fact, even with this happening, our world has managed to outshine the small shadow that has been cast by this impediment. Actually, and in all seriousness, I have never been happier in my life. So you can imagine how I might feel if presented with a succulent homemade apple and blackberry pie; a sumptuous chocolate pudding; or maybe even a sweet and delectable rhubarb crumble?! We all dream of that heaven on Earth don't we?

Annie is the sole reason I have been able to complete the writing of this book. She is also the reason I have been able to affectively and productively process my contact experiences and to discover the hidden strength and wisdom in them and within me. Annie has been the foundation beneath my transformation; an unfaltering source of astuteness, good judgment, acumen and incredible intuitive intelligence that has guided and nurtured my journey through this maze of dead ends. Without her unconditional love I would never have been able to make the mistakes I have, only to then get back on my feet, dust myself off and continue on with the job. I just could not have done it without her.

Annie is a multifaceted person. Her life journey has been extremely complex and not without its tragedy and trauma. At times she has worn many different hats: art school administrator, bank clerk, continental au pair, croupier at Hugh Hefner's London Playboy Bunny Club, all-round health therapist, hypnotherapist, past-life regressionist, life counsellor, artist, dress maker, metaphysical investigator, author, cordon bleu chef (with certain skills missing of course!), full time mother and house frau, painter and decorator, jewellery maker, garden designer, nutritional expert, teacher and student. If I have left anything out – which I probably have – she will gently remind me.

She has made a positive impression with everybody she has met, established deep and meaningful life-long friendships with a host of incredible individuals all over the world and tragically, sadly and unavoidably lost others whilst on that life journey.

People *remember* Annie. Through a few heartfelt words or a simple act of kindness, the affect of her contact remains. Meeting Annie has been known to change one's life; and there will be a host of people who will bear witness to that fact, but none more than me.

I have been married before, so I was a little wary about introducing Annie to my family, as you never know how people – even cherished loved ones – will react to newcomers. I should never have had any reservations though because with Annie being Annie everything worked out fine. My two sisters have become the sisters that Annie never had and they could not be closer with her. Their very young children – my nieces – have become, for Annie, the temporary grandchildren that are yet to arrive. The girls acknowledge Annie as the remarkable and special human being that she is without really knowing or understanding why, except for how she talks and interacts with them. When we visit they rush to the door squealing her name in excitement and anticipation for the experience that is to follow. The only trouble is that poor Annie always returns home covered in tiny bruises from them enthusiastically clambering all over her, cuddling, checking out her homemade jewellery and redesigning her hair. It is truly something to behold. Getting them to calm down and off to bed after we have left is quite a feat I am told.

As for the shallowness of physical beauty, Annie continues to increase my pulse rate and body temperature every time she walks into the room. Being a member of the male species we have been known to be creatures of whimsy and our interest in a lot of things can wane depending on which way the wind is blowing. Gratefully, my female genes have always been dominant in the field of appreciating the finer things in life so my enjoyment of Annie's physical beauty has only increased the longer we have been together. I am a renaissance man and the Rubenesque qualities of her appearance bring out not only the finesse of my character, but the animal also. My love for this woman burns with a passion that sets my heart ablaze, the light of which is recognised for its truth. Annie is my last thought each night before I am embraced into the arms of Morpheus and the first thing I think of before I even open my eyes to greet the day. She fulfills me and feeds my hunger. She satisfies my longings and makes me crave for more. She is the woman that I always dreamt of sharing my life with. She is fantasy and reality. She is *Annie*.

My wife's knowledge of the world is immeasurable and, at times, unbelievable. Up until just the other day I was yet to come up with a word that she did not know the meaning of or know how to spell. With any subject that you might like to name – within reason – Annie has acquired at least some level of knowledge or experience through reading, study or a hands-on encounter. How she has been able to do this in a relatively short life I just don't know, but she has. When discussing a particular subject with a trained specialist they are always surprised to find that my smarty-pants wife actually understands what they are talking about and able to join in with the conversation. And she's never bigheaded about it, always low-key and humble, always matter of fact. That's one of the reasons why people love her; *Annie is real.*

By now, dear reader, you would have realised that the path I have trodden has not always been stone free. It would be fair to say that my family background has been far from unique, either in my early years or my young adult life; divorce and everything that comes with it is sadly a normal feature of our modern world. Therefore the challenges that arise from heartache can ripple through into later years and it takes time, wisdom and love to calm those troubled waters. Clearly the other more exotic aspects of my life have also played a part in being able to establish a stability that those around me could be comfortable with.

I met Annie when the winds of change were probably blowing their hardest and she has remained by my side – steadfast and true – intuitively knowing in her heart that I would eventually find safe harbour in our shared love. With that secure mooring came new and exciting features to our life together: one being a playfully mischievous aspect to Annie's character which has recently begun to emerge and with that has come a *shared* laughter, jollied along with a humourous script and funny voices. I gratefully acknowledge that Annie now feels comfortable enough to explore this role because of the new security that she feels in our life together.

The phrase 'as sharp as a knife' could have been written to describe my wife, whose analytical brain enables such clear perception of a situation that she has formed a plan and moved on whilst the rest of us are still scratching our heads. Obviously with any picture it is made up of a wide-ranging palette of colours and Annie also has a spectrum of character features that seem to contradict others: her malapropisms are famous and once diagnosed leaves one in stitches of laughter; also her aptitude to begin a sentence half way through because she has already 'spoken' the first half in her head. These are just a couple of facets of a mind that is always running at double speed, spinning too many plates and juggling too many balls. But we would not have it any other way; because in my experience one particular behaviour is not likely to come without the other and I find every single aspect of my wife's complex make-up seductively endearing.

The word fragile would probably not be one commonly used to describe my wife, but it would be true to say that she possesses a delicate nature. People have confused her softness to be a weakness in the past, but have learnt to the contrary when they have discovered that she is indeed a strong and determined person who will not be taken advantage of. She is a very determined individual – *it might even be called stubbornness* – but she is not likely to be diverted from her path if her heart believes the course to be true. And when it comes to injustice there is nothing short of a brick wall that will stop this super-powered woman from addressing the imbalance of the situation and seeking retribution for the wrongdoing. One might witness tears in her response to an injustice, but be aware that they will be backed up with a force to be reckoned with. In that respect and in a million other ways, I am proud to be Annie's companion on life's journey.

I love this woman with every fibre of my being and begrudge every second that I have to be apart from her (which is, thankfully, very few). Annie *gives* to everybody, but has always found it difficult to *accept* on any level; therefore I hope that through my love and devotion I have helped her to discover a self-awareness of who she really is and to be comfortable and accepting with that knowledge. I would also like to believe that Annie is now free from anxiety in her own skin, for she is truly beautiful in every sense of the word and the reflection in a thousand pair of eyes confirm it to be so. If I have somehow played a part in that overall transformation, it is in small part to rebalance everything that she has helped me to achieve. I will be forever in her debt ... and it is such a lovely place to be.

So, in conclusion, I will say this: *intelligent, playful, inexhaustible, versatile, smart, inquisitive, loving, humourous, creative, passionate, generous, 'hot', kind and caring, relentless, intuitive, elegant, warm, understanding, supportive, gorgeous, determined, stylish, funny...* yes, but can she make chocolate fudge brownies? Actually she can. Through perseverance and self belief she can now successfully make desserts. She still doesn't like ice cream though!

*"Illegal aliens have always been a problem in the Unites States...
ask any Native American."*

Robert Orben

Social Event: Our Beloved New York City

New York City – May 2003

The day after arriving in New York City, one of the first things Annie and I did was to find an authentic little New York restaurant in Greenwich Village to have breakfast. Walking west we found such a place on the extremities of The Village, just before you reach the Hudson River. I can't remember the name of it now but the interior was surprising; being padded out in old, dark wood with a plethora of antique fishing regalia: old fashioned rods and nets, weathered wicker tackle baskets, sepia toned photographs of memorable catches and endless shelves of dusty, leather-bound books. We ate enough to ensure that we wouldn't have to eat again until the evening – this was something we quickly discovered about eating out in the Big Apple; the portions are much bigger than anywhere else in the world.

Unfortunately, just before we had left England I had severely stubbed my big toe. By the way it felt on the first evening in The Village as we sat outside a bar having our first drink in town, I knew it was probably broken. Now, there's not much one can do about a broken toe except suffer it. The only problem was that I didn't have any open toe sandals to wear, only close fitting sports trainers, new ones at that and not even worn in yet. So I had a good idea, I would wear Annie's flip-flops and take it easy until my injury began to heal. Unfortunately, because of having unbearably sensitive skin, within an hour of wearing them I also acquired huge blisters from the rubbing in between of my toes. Ahhhh, jeeeeezzzz–looouiisssseeee!!!! as they say hereabouts.

After breakfast, we jumped in a yellow cab and made our way down to the Pier, which sits in the shadow of the Brooklyn Bridge. Our plan was to purchase some sort of open toe footwear for me to wear that would be gentler on my sensitive little tootsies. The Pier area is a major tourist attraction and has a shopping mall and also an old sailing vessel that sits majestically anchored in the East River. We made our purchase and I hobbled my way up to the Brooklyn Bridge, where we stood taking in the panoramic views of the city behind us and Brooklyn just on the other side of the bridge.

There was a very important landmark that we had definitely come to see. In Budd Hopkins book, *Witnessed: The True Story of the Brooklyn Bridge UFO Abductions*, it tells a story that began on 30th November, 1989 at 3 a.m.. On the Lower East Side of Manhattan a motorcade carrying international diplomats, including a U.N. representative of the highest standing, breaks down and comes to a halt beneath the FDR Drive. An enormous reddish-orange UFO hovers above an apartment building, where a pretty young woman is seen by the members of the motorcade to float out of a 12th-storey window, accompanied by three small grey aliens, and into the UFO. Because those involved, including the floating woman (Linda Cortile), the U.N. diplomat and his security men (probably "CIA or NSA"), eventually read Hopkins, bestselling book, *Intruders*, they all end up contacting him. Without a shadow of a doubt, the 'Brooklyn Bridge incident' is the most witnessed UFO abduction of all time – with new witnesses *still* coming forward to this day. Photographs contained within the book show a very clear depiction of Linda's apartment building and the very recognizable water tower that sits atop it. The apartment block is one of a few buildings, all exactly the same, that can be clearly seen from certain vantage points on the Brooklyn Bridge. Since reading Budd's seminal work on this abduction case, both Annie and I had dreamt of possibly meeting up with Linda on our visit to America, because we were so touched and affected by her story, especially the determination and courage she showed in dealing with the actual event, but more importantly the aftermath that continues to impact upon her world. As we stood on the bridge determining exactly what was her building, we imagined her looking back at these two typical tourists desperately trying to fit in with the hustle and bustle of New York City life. Within just a few days our dreams would be fulfilled.

During that first week in Manhattan, we attempted to soak up as much of the traditional New York City as possible; we went to the breathtaking Statue of Liberty (only just re-opened after the barbaric atrocity of 2001), the historic Ellis Island, the iconic Empire State Building, the buoyant Circle Line River Cruise, the Met, the Guggenheim, the Natural History Museum, and out to Queens for the temporary housing of MOMA, beautifully air-conditioned movie houses on 42nd Street, the expensive, yet unmissable Broadway theatres, an unforgettable Soul-Food breakfast in Harlem, and shopping, shopping, shopping – here, there and everywhere. We walked our little feet off – looking back we can't imagine how many miles we must have walked that first visit to New York; I don't think we could do it now (although we'd have a bloody good try!).

Before I had left England, I had been in contact with the Intruders Foundation in New York City attempting to make contact with Budd Hopkins to arrange a meeting whilst in town. The Intruders Foundation is a non-profit making organisation that investigates the world of Alien Abduction and Budd was its creator and the heart and soul of its infrastructure. Not having access to an Internet link, I was unable to check my emails on a daily basis whilst in New York, but did so eventually after being in town for just under a week. I hadn't really expected to hear from Budd, not because of a lack of professionalism but because I felt that my request was unimportant compared with the other work he must be involved in. Apart from being an established and successful artist, he is also an authorative author and investigator of the phenomenon that is known as Alien Abduction. I just saw myself as a small fish in a very large pool. Therefore, it came as quite a surprise when I discovered several messages from Budd asking me to contact him as soon as possible. I eventually spoke on the telephone with him and he invited me to his home in the Chelsea district for a meeting.

On the morning of our appointment I left our apartment at 9 a.m. and walked up the Avenue of the Americas to Budd's home and studio. I found it fascinating walking along 6th Avenue with the busy throng that was New Yorkers on their way to work. It really made me feel like a 'homey'. When I rang the bell and the door was opened I was greeted by a tall sophisticated-looking gentleman in his 60s, who warmly invited me over the threshold. I was directed up the staircase to a sitting room occupied by a substantial and impressive library, a magnificent art collection and two exquisite cats. The latter curled around my legs and followed me to the couch as I sat down. I welcomed their attention as I sat – just a little nervously – enjoying the warm hospitality that I was being afforded. Our conversation was general to begin with, as Budd asked questions about our trip so far and my thoughts and opinions of New York City.

Budd is a tall, slim, debonair man with a good head of silver hair and a twinkle in his eye. After being struck down by polio during the pandemic of the 1930s, the effects of the crippling disease still lingers and causes him to limp slightly, but does not appear to hold him back as he lives his life with a distinct vibrancy and enthusiasm. He is also a nationally known Abstract Expressionist painter with works in the collections of the Guggenheim, Whitney and the Metropolitan, as well as Boston's Museum of Fine Arts and New York's Museum of Modern Art. Budd has been the world's leading authority on the subject of Alien Abduction for nearly 40 years and has authored such decisive and shaping works as *Missing Time* and the New York Times bestseller *Intruders*, which was made into a CBS mini series. He was the inspiration for the Harvard psychiatrist, Dr John Mack, to take up the gauntlet of bringing a more scientific investigation into the subject of UFOs, Alien Abduction and contact with extraterrestrial life. I was in no doubt that I had the profound honour of working with the crème de la crème.

What Budd helped me to do over the next two days was priceless as far as I am concerned. Not only did he help me to begin the journey that would eventually lead to that place where I would no longer fear my contact with these extraterrestrial intelligences, but he would play an important part in bringing all of my past experiences into a much sharper focus and enable a more profound recall of my ongoing exchanges. All of this would ultimately enable me to transform from being a powerless Abductee to becoming a self-empowered Participator in the events that have played such an important role in my life.

On my last afternoon at Budd's home, he invited me to attend a meeting of the Intruders Foundation, an ongoing support and advisory group, which came together during the Autumn and Spring of each year. Each time the group met a high profile invited speaker would attend and conduct a presentation based on some aspect of the UFO phenomenon. On that coming Saturday, after the main speaker, Budd asked me if I would briefly outline my history to the assembled group. It felt like a great honour and of course I accepted.

As I thanked Budd profusely for his help and nurture over the past two days I enquired if Linda Cortile would be there on Saturday at the meeting. He assured me that Linda never misses a gathering of the Intruders Foundation and that he would introduce both Annie and myself to the lady in question. Our remaining time in New York City literally flashed by, although we were so excited about culminating our trip to New York City with a visit to Budd's group on the final night before we flew home.

So, on Saturday 10th May, we arrived at the Intruders Foundation in good time and were greeted by Budd and shown to our seats. The presenter that evening was a gentleman called Frank, who had an amazing tale to tell of his ongoing abduction experiences. His presentation lasted about sixty minutes and included some very interesting video-taped evidence. We had been looking out for Linda's arrival and had spotted somebody come in just before the evening's event begun. She had her head held down and disappeared into a seat at the back of the room. Based on her low profile entrance I wondered if she would want to meet with a couple of strangers but I was very pleased when Budd brought her over during the interval to say hello. My hopes were somewhat dashed when she held out her hand, briefly exchanged pleasantries, then vanished once more into the shadows at the rear of the room.

When Frank completed his presentation, Budd introduced me to the gathering and invited me to take the microphone to talk informally about my experiences. Once finished I handed the microphone back to our host and he covered the closing proceedings for the night. As Annie and I hovered, not sure what to do next, Linda approached once more; she had transformed from the shy, retiring woman whom we had seen arrive a couple of hours earlier to the gregarious, warm and inviting woman that now engaged us. I would later come to understand Linda's need to appear reticent when first meeting people, based on some unpleasant experiences with certain debunking individuals since her abduction encounters. She enquired if we had plans and asked if we would like to join her for coffee. We had no plans and suggested that she travel down to Greenwich Village with us, which was sort of on the way home for her and then we could have a drink or maybe a bite to eat. Linda told us that she knew The Village extremely well and suggested an authentic Italian restaurant that she had frequented as a young woman on the junction of MacDougal and Bleecker Street, which was in fact about fifty yards from where we were staying.

As our yellow cab prowled slowly down Bleecker Street in the still busy, late night Saturday traffic, I watched the patchwork quilt of people strolling and chilling. I have never been bored of that sight and believe I never will.

The Café Reggio, MacDougal Street is at the junction of Bleecker Street, Greenwich Village, NYC. It is pure Italian my friend and not to be missed. It must have been at least 10 p.m. when the three of us perched around a small metal table outside the café. The temperature on that warm May evening was extremely pleasant and enabled us to sit very comfortably sipping on our selection of coffees, cognacs and Linda's Italian version of Coke.

As for Linda, she is a very classy lady; an attractive and smartly dressed woman who looks much younger than her years. One could not hope to meet a more generous and kind person, with a sensitive nature that can often cause her distress when she does not see it reflected in others. Her accent and dialogue is authentic Italian Lower East Side Manhattan and is recognizable from a million New York movies. Although New York City has shifted in its attitude over the past two decades, Linda has remained authentic to her Italian working class roots and reflects that through her home life as a hard working wife, mother and grandmother. Dear, sweet Linda has a plethora of stories and anecdotes to tell about her life in The Big Apple (which is a book on its own and hopefully for her to tell one day) and we have spent many long hours with her being regaled by stories of life amongst the more colourful characters thereabouts, including an encounter that

Linda and her husband, Steve, had with the notorious Mafia boss, John Gotti, where he came off worse for wear – I kid you not!

To have the most informed view and appreciation of who Linda Cortile really is, it is vital that you are one of her very few close friends and confidantes without that insight it's not that easy to become as close to her as you might want. Linda is, without doubt, one of our dearest and most special friends in the world and I believe that she feels the same way about us – and for that we are truly blessed. Without her gift of friendship, the next best thing to do is to read Budd Hopkins's book *Witnessed*, which tells the harrowing tale of her alien abduction and subsequent spell-binding story of high intrigue that is one of those really important movies that have never been made, but should have been.

Our talk was immediately easy on that night and we would later comment on how comfortable we all felt right away. Annie and I had a million unanswered questions about Linda's case and we unashamedly set about asking them, which Linda was extremely generous in answering with obvious honesty and delightful eloquence. There was clearly so much more to her abduction story than had been written about so far and we were to learn exciting new aspects that night and throughout the ongoing years of our friendship. If and when Linda decides to publish her story, you are all going to be literally blown away by what she has to tell you, because, as far as Annie and I are concerned it is one of the most important stories for the rest of human kind – once more I say, I kid you not!

After a while Linda wanted to know the ins and outs of my contact experience, so where better a place to start than my story about Little Star and Peter Pan. When I reached the moment where the longhaired, pale, skinny teenager – now referred to as Peter Pan – calls the tiny, cheeky, imp-like child Little Star, Linda's jaw visibly dropped. Annie attempted to interject at this point, but I gestured to her to hold-that-thought, as I wanted to observe how Linda's response would play out. Her apparent look of pure incredulity stayed firmly fixed to her face. "What is it," I asked her?

"You have just awoken a memory for me; something that I have never shared before, not even in my sessions with Budd." (Linda had had some regressional hypnosis sessions with Budd Hopkins to access her memories concerning her numerous alien abduction experiences throughout her life.) "This is something that even I had forgotten, but by you just saying those words it's just sprung back into my memory".

"What words Linda?"

"Little Star."

Although Linda was looking directly at me, I could tell by the glassy, faraway look in her eyes that her mind was elsewhere. She was clearly recalling something that had been stored in a very deep recess of her mind.

"I can remember during one of my abductions... it might have been more than once actually," Linda pondered. "I heard somebody say the words Little Star. This voice was behind me, so I couldn't see them, but they called out that name Little Star. It came up when Budd worked with me, but for some reason or other, I didn't tell him." Linda looked quizzical. "That's really strange isn't it, that I didn't tell him?"

"Umm," I agreed softly. I couldn't believe what I was hearing and by the look on Annie's face, she felt exactly the same.

For the next thirty minutes or so, Linda didn't really feel present although she did respond in the correct places as I proceeded to tell bring her up to date with regard to my contact with Little Star. Then, completely unexpectedly and out of nowhere, Linda sprang back into the conversation with an animated burst.

"There's something else about Little Star that I remember."

I stopped in mid sentence, happy to allow her to continue.

"I have a best friend, her name is Gloria (pseudonym) and she's an abductee too." Her eyes flashed with excitement, as if she had been trawling her memories for a missing piece of the puzzle and she had now finally found it. "I remember sitting in her car. She had this sort of plastic star thing hanging from her rear view mirror. It used to really annoy me how it would swing backwards and forwards distractingly as we drove along. Anyway, I asked her what it was and she said, 'It's my little star,' and gesturing towards the roof of the car, pointing up, she then said, 'It's what they call me, their little star".

It was now my turn to open my mouth and leave it there gaping in amazement, as I was totally and utterly flabbergasted.

"How old is she?'

"About my age," Linda replied vaguely.

Based on my mathematics, that excluded Gloria automatically from being my 'Little Star'.

Linda obviously picked up on my train of thought. "No, I'm not suggesting that she's the girl in your experiences." She shook her head. "No, I never thought that... at all. But..." she trailed off.

"What can it possibly mean?" asked Annie.

I felt so exhilarated. This just brought such integrity to my contact; not that I disbelieved my own encounters but it made me realise that I had been looking at the whole contact experience through a very narrow aperture. I knew that I was just a piece in a very complicated puzzle, but I now felt connected to the bigger picture.

We all exhaled together, releasing, I believe, a breath we had all been holding in unison. Like a tender, warm embrace our mutual response to this revelation just seemed to flow through an invisible bond that held us all in that moment.

The waitress came to our table and enquired if we required another drink. We sat there waiting in comfortable silence as our selection of drinks was placed before us. Stirring sugar into my coffee I sat pondering on how to proceed. "This is going to take some time to absorb, isn't it?"

Annie and Linda agreed.

"I would really like to go and telephone Budd and let him know," I enthused.

"He'll be fast asleep. He's got an early flight in the morning," Annie reminded me.

After a little while, Linda looked at her watch and was amazed to see that it was now nearly 1.30 a.m. It was Mother's Day tomorrow, she told us, and she had the whole family coming over for lunch. We walked her up to the junction of Bleecker and 6th and I hailed a yellow cab. My last memory of that night is seeing her beaming face smiling through the rear window of the cab as she waved frantically. To paraphrase Mr. Humphrey Bogart in the movie *Casablanca*: 'This was the start of a beautiful friendship'.

"When the long awaited solution to the UFO problem comes,
I believe that it will prove to be not merely
the next small step in the march of science,
but a mighty and totally unexpected quantum leap."

Dr. J. Allen Hynek

Social Event: Our First Trip to Laughlin

Laughlin, Nevada – February 2004

Towards the end of 2003, after reading the advert in a copy of the UK edition of *UFO Magazine*, Annie kept suggesting that we should attend the UFO Congress Conference in Laughlin, Nevada the following February. Apart from the point that it would cost a lot of money to carry out this trip, I was still a little reluctant to embrace the larger structure of the UFO community. Why this remained so, I had absolutely no idea. Clearly I had dipped my toe in the pond by attending the UFO conference in Leeds back in November of 2001, but because that had opened some sort of Pandora's box for me with regard to my own experience memory I was still feeling a little overwhelmed.

Now, Annie has a way of sensing that something is right, without necessarily being able to back up her intuition with any firm facts. With regards to going to America to take in this conference, she just said that essentially it 'felt right'. In fact, the more I placed obstacles in the way, the more she became really quite impassioned that we really had to go without argument. Finally acknowledging how convinced she appeared to be about this matter, I agreed to go. Thankfully, as the years have progressed, I now listen and respond positively to my wife's intuition sooner rather than later. Why do 'we men' take so long to wise up?

This particular conference is a 'must do' experience for all those people with an open and enquiring mind. Not only does it apply itself to the subject matter of UFOs and Alien Contact but also a wealth of subjects ranging across the board of new science, paranormal activity, astronomy, astrophysics, exotic technology, hidden world history, hidden energy technology, conspiracies, crop circles and many others. All of this is presented by an ever-changing team of world-class lecturers and presenters and supported by a high-class team of technical wizards. Up to 1,000 people from all over the world come to the Nevada desert every year to have their brains educated and their minds expanded for seven days. And I was reluctant at first to go?!?!?!

Initially we didn't actually know anybody, although the crowd is a very friendly one and you can so easily get into conversation between presentations and at meal times. On Wednesday, after three days there, Budd Hopkins arrived and we met up with him at the 'Meet The Speakers Dinner' in the evening. It was so nice to see Budd again after nearly a year and with that reunion came an open door to a wealth of new friends. Travelling with Budd that year was Leslie Kean, a world-renowned investigative journalist and Director of Investigations for The Coalition for Freedom of Information. Leslie is an extremely 'smart cookie', as the saying goes, and continues to battle the system to release the truth behind the UFO secrecy and, along with Budd, has now become a very dear friend of ours. Also in Budd's company that year was John Franklyn, who is an established author and UFO investigator and Anne Cuvelier, who remains the vital orchestrator attracting new individuals into the orbit of other like minds and continues to expand our universe of contact with each other; without her support I would not have achieved what I have over the last few years. Through these exciting new contacts we had, by the end of the week, established a whole new network of friends and contacts that remain vitally in place to this day.

Annie and I had attended the Abductees Support Group, held each evening after the day's presentations and administered that year by Mary Rodwell (Alien & UFO Researcher and Principal of the Australian Close Encounter Resource Network [ACERN]). We were both rather naive at that time with regards to the diverse elements that make up the UFO community, so it came as quite a surprise when we were confronted by some of the more wacky characters from the arena. To begin with we found it rather amusing to hear some of the rather extreme stories of alleged contact which would have fitted more comfortably in to a science fiction B movie of the 1950s, but eventually they became rather irritating as I realised that the 'smaller' voices, the ones that really needed support from the overall group, were being suppressed and not being allowed to speak because of the more selfish and dominant factors within the room. Therefore, because Mary and I were now friends, we concocted a plan for the second evening to head-off the aggressive participants and free up the space for the more timid, but more deserving voices to be heard. It wasn't easy, but we achieved our goal. And to that end, many people who had never had the opportunity to speak of their contact experiences before got the chance, not only to express themselves, but also more importantly, secure support and nurture from others in attendance.

On the morning of Thursday about half a dozen of us were invited to speak at the 'Experiencer's Open Mic' session. This event gave those who wanted to participate the opportunity of speaking to a closed room of attendees without the inclusion of the media or having the presentation recorded, thus ensuring their anonymity, to a certain extent, if so desired.

As I sat in line with my fellow presenters, I became aware of the person to my left and sensed their apprehension of sharing the details of their contact experience. Turning to introduce myself and expecting to find somebody of reduced stature who would reflect the uneasiness I was sensing, I was surprised to discover a 'Redwood tree' of a man, tall and bold in his seat and with no *obvious* sign of uneasiness. With a European flavoured accent he introduced himself as 'Donald' and quickly confirmed my suspicion by telling me he was on the verge of stepping out of line. For some reason or other I took it upon myself to encourage his continued participation in the events that were to follow. In fact my encouragement went as far as physically assisting him towards the microphone when his turn arrived. Donald moved towards the stage with a committed step and went on to share an amazing record of his contact experience with great eloquence.

After the events of the morning were over, Donald came over to thank me for helping him to make his mind up. He truly believed it had been a turning point and that he would now continue to explore the mysterious memories that continued to cloud his perfect recall of some aspects of his childhood. In fact this is exactly what has now happened with Donald taking the opportunity to pierce the veil that was stopping him from remembering all the details of his contact experiences.

Thankfully we have been blessed with the ongoing good friendship of Donald and besides sharing his company at other conferences, we have also spent time together at the house of our mutual friend Anne Cuvelier. I still experience Donald as I did upon our first contact: intelligent, generous, kind, caring and remarkably eloquent – and all in Redwood size portions.

When it had been my turn at the 'Open Mic' I had spoken about my 'David Bowie Night' experience and the audience very warmly received it and I guess it was probably at this point that I decided to begin talking properly about my experiences. It would be quite a while though before I worked out in my own mind why I should be doing it, therefore I had to rely temporarily on Annie's intuition that it was the right thing to do.

All in all, attending the conference was one hell of an experience and we topped it off by hiring a car and travelling down to the Grand Canyon after we finished, then on to the red-rock paradise of Sedona for a couple of days. My ever-lasting memory of this particular part of the trip is Annie and I lying in a hot tub on the veranda of our hotel room, late at night and looking up at the vast black velvet sky sprinkled with sparkling gems and 'wishing' for a UFO to fly over! Sadly, on that night anyway, it never happened.

On our ride back through the desert to Las Vegas to catch our flight home, it was extremely hard to maintain a steady 55 miles per hour and not break the speed limit. Therefore, it was inevitable that eventually I would stray into that part of the speedometer that makes police officers appear out of nowhere. As I came over the brow of a hill at... let's just say I was exceeding the speed limit... there in a previously unseen dip of the landscape was a police car just waiting to make my acquaintance. As the police car's siren wailed and the blue lights flashed I slowed down to the side of the road, trying very hard to remember that in the USA one stays in the car and waits to be approached, unlike being apprehended in the United Kingdom. A young policewoman asked to see my licence and insurance and I attempted, rather pathetically, to 'schmooze' my way out of trouble. As I reached behind the driver's seat for my documents, the police officer's hand went automatically for the handle of her gun, obviously anticipating something other than my driving licence to be produced. Now, I could say that she relaxed her stance and responded positively to my extremely slow production of what had been requested, but I'm going to give you the conclusion to the story that I took home with me and still dine out on to this day.

As I reached behind my chair, fumbling for my travel bag, with her right hand the burly police officer pulled her revolver from its holster and placed the barrel of it against my temple. I was then pulled aggressively from my car and flung unceremoniously across its hood and pistol whipped into submission. Well, that's how I like to tell it anyway. I know, I know, too many Dirty Harry movies!

"Each time we face our fear,
we gain strength, courage, and confidence in the doing."

Steven Jones 2010

Experience: Interrupted Mail
and Black Helicopters

Essex, UK – February 2004

When Annie and I returned from our magnificent trip to the UFO Congress Conference in Laughlin, Nevada in February of 2004, we were both feeling tremendously excited and happy.

The years since 2001 had been such an emotional rollercoaster for both of us, due to the fact that it was only then that I really began to accept and finally embrace who I really was and what was happening to me. By meeting and connecting with such an amazing group of new friends from within the American UFO community, we had created a bedrock of support that would play a vital role as I continued to evolve and grow on this new path of self-empowerment. They not only opened up their homes to us on subsequent trips to the United States but would also offer a degree of succour that continues to amaze us both. As our friendship has grown their shared wisdom and nurture has only been superseded by their unconditional love and care for us. We hope, that in some small way, Annie and I have been able to express our gratitude to this group of truly outstanding human beings and somehow reflect back their feelings and actions.

In the weeks after our return we began to notice that our mail was arriving already opened. It wasn't that they appeared tampered with and resealed, they had been completely ripped open with no attempt to conceal the fact! At first we thought that there might be

221

some sort of logical explanation – well you do, don't you – but then, when it persisted we resigned ourselves to the fact that our post was being interfered with. But by whom? And more importantly, for what reason? We postulated that it might be something to do with the fact that we had been to the UFO conference and that we were now having regular e-mail contact with some extremely high profile people in the field. Would that be enough? Were the 'powers that be' concerned enough about the likes of us to want to read our mail or maybe it was just an intimidation tactic?

Additionally, we started to have peculiar experiences with our telephone line: sometimes when we picked up the receiver to make a call there wouldn't be a dial tone, only silence, punctuated with whispered voices and strange noises. Also, when we would be in the middle of a conversation we could hear another voice in the background, but we could never make out what they were saying. We were able to establish that there wasn't a fault on our line and that we were not getting crossed-lines, but we were never able to pinpoint what the problem was. It still happens every now and again, but it has never actually bothered us; in fact, we have often played games with whoever was listening in by saying things like "…if you want to know about UFOs and what we're up to, just pop round for a chat and we'll tell you everything that we know!"

Although Annie and I have a passing interest in the plethora of conspiracy theories that do the circuit these days, our curiosity is healthy and realistic. We are certainly not paranoid people and consider ourselves to have our feet firmly planted on the ground. Therefore, when the next aspect of apparent surveillance presented itself we could not help but find it all extremely amusing and ridiculous at the same time.

At the rear of our house, we have a detached building that has been converted into an art studio and whilst working on a warm summer evening in July of 2004, I was to encounter a very strange occurrence.

During the process of preparing a canvas, the venetian blinds pulled down to block the rays of the setting sun and the windows slightly ajar to allow the flow of fresh air and the soothing sounds of our garden, I was momentarily distracted by a very low distant hum.

The far-off noise was hardly loud enough to penetrate the haven of the artistic environment; therefore I remained focused in the moment and continued to work on. Gradually, I became more and more aware of what was going on outside than inside. The noise was now invading the space with a deep repeating resonance. I recognised the pulse as it reverberated through the air. Chomp, chomp, chomp. The noise continued. It felt like it was right in the room, throbbing up through the floor and vibrating into the chair I sat on. Whop, whop, whop. Whatever was generating this intrusion was very, very close.

Frustrated, I jumped up, pushed a couple of blind slats aside and closed both of the windows. Thomp, thomp, thomp. It made virtually no difference to the auditory intrusion. The easily recognizable sound of the rotor blades of a low-flying helicopter beat just above the roof of the room where I stood.

When I went back into the house, I found Annie out on the doorstep with her eyes turned skyward and surveying all around.

"What was all that about?" I enquired.

"Did you hear it too?"

I laughed. "What do you think? I reckon the whole estate heard. Did you see what it was?"

"It was unbelievable. I was watching TV and I heard this deep pounding noise right off in the distance. It got closer and closer until I realised it was the sound of a helicopter. It became so intrusive that I went outside into the street to see exactly what it was." Annie shook her head in disbelief. "Right at the bottom of the road, hovering relatively low in the sky was a big, black helicopter. It just sat there, not moving."

"What did it do next, because it sounded like it was right overhead?"

"It hovered for about five minutes I suppose, then it began to get lower and lower, until finally it couldn't have been more than 20 or 30 feet above the roof tops. Then the really weird thing happened: it just sort of drove down the road – in the sky still of course – but it just seemed to use the road as a guideline. When it reached us it manoeuvered a little bit until it was directly over the studio and then it just sat there. Everybody came out of their houses to watch, all the kids were running up and down – it was completely mad!"

Annie's look changed to one of puzzlement and she shook her head once more. "The other peculiar thing was that, yes, it was noisy, but not as noisy as you might expect it to be. It was so unbelievably low and so colossal, but it sounded sort of... muffled in a way... I can't really describe it."

"Well, it sounded loud enough to me."

"Sure, it was loud, but... sort of strange at the same time." Annie now had a crooked smile. "Anyway, not long before you came out it just shot straight up – that was really fast – and it disappeared over the rooftops towards the motorway. I kept looking and that's when you came out."

"Did it have any letters or numbers on it?"

Annie thought for a second.

"No, now you come to mention it, no, there wasn't anything." Her eyes looked around to the bottom of the street.

"It was black, but that was odd too: it wasn't shiny, it was a matt black, but the surface seemed a bit like glass; as if it was translucent in a way."

'Okay, that's interesting."

I thought for a moment.

"I'm going to document all of this before I forget. Anything else?'

"I think that's more than enough, don't you?"

"Then one becomes two
and all of a sudden we are of a like-mind.
That's when the changes begin."

Steven Jones 2010

Social Event: A Weekend in Connecticut

Connecticut – May 2004

The year 2004 turned out to be a rather busy one for Annie and me: we went to America three times – once in February to attend the Laughlin UFO Conference, then again in May to spend time in New York with Debbie Kauble and to make a presentation at Budd Hopkin's Intruders Foundation, and finally we were invited to Anne Cuvelier's home in Newport in August. During our three week vacation to New York City in May, we also went up to Connecticut for the weekend to spend time with our very good friends, John and June Franklyn.

Their aim was to create a new gathering of UFO enthusiasts, similar in style to Anne Cuvelier's annual get-together in Rhode Island (to be explained in detail in another part of the book). Not having the extended accommodation that she has, several of John and June's invited guests (including Annie and myself) stayed in a very comfortable local hotel, just ten minutes up the road.

In attendance, as well as Annie and me, would be: Peter Robbins – the co-author of the famous UFO book, *Left At East Gate*, published in 1997, which told the first-hand account of the 1980 UFO incident at the American Air Force base Bentwaters/Woodbridge on the east coast of England, the attempt by the military to cover it up, and the independent in-depth investigation that was eventually carried out. After having read the book when it was first published, and acknowledging the important role it played on so many diverse levels we were both very excited at the prospect of meeting Peter. What we were to discover was a man who is totally committed to his unceasing pursuit of the truth and his exceptional ability to retain every single fact regarding his research!

Carol Rainey, Budd Hopkin's wife at the time and co-author of the seminal work, *Sight Unseen*, was going to be there too after travelling up from New York City with Peter. We had already met Carol in New York the previous year whilst I was there to work with Budd, and we had also recently seen her carry out a joint presentation with her husband at the UFO Congress Conference in February. The opportunity to get to know her better had not presented itself as yet, and based on her substantial contribution to the book *Sight Unseen* we were really looking forward to talking further with this extremely intelligent woman.

The final guests would be the twins Jack and Jim Weiner and Charlie Folz, three of the main characters in Raymond E. Fowler's book *The Allagash Abductions*, which tells their story of alien contact and abduction in 1976, deep in the wilderness that is the Allagash Waterway of Northern Maine. Annie and I had only read the book in the previous six months and had been totally engrossed in the phenomenal tale of their experience and moved by the details of their emotional encounter. Additionally, there was Mary, Jack's wife, whom we found to be a very sweet and funny lady, whose favourite expression was the whimsical saying 'What eeevvveeeeer!'

The party gathered, with Carol and Peter still missing by the time we sat down for our evening meal. After a telephone call from them we were to discover that they had gone past their connecting rail station after being so absorbed in conversation. Once we had spent time with Peter the next day we found out the reason why: he is a very passionate man when it comes to most things in life, but when it has anything to do with UFOs Peter can simply go without food, drink and, obviously, connecting rail stations. Carol later explained that dear Peter had not really taken a breath since leaving Grand Central in New York City, so it came as no surprise to find out that they should have alighted from their train a few stations back. That, Peter explained later, is one of the results of being so immersed in one's subject matter.

When we reconvened the next morning, sitting out in the beautiful garden that surrounds the house, I felt extremely honoured to be able to have a direct, one-to-one conversation with Peter about the subject of his book. By lunchtime he had shared a wealth of additional information about the case and been able to explain, in even greater detail, the facts surrounding this amazing story. If you have not read his book yet, I do advise you all to do so. Additionally, if you are ever able to visit the site in England where it all played out, please do not miss the opportunity. It was an experience that Annie and I had a couple of years later, when we traveled there in the company of our new friend, Peter Robbins, and not only stayed at the bed and breakfast he resided at when he first carried out the investigation for the book, but we also had a moment-by-moment guide by Peter through the forest where the event actually occurred. An event of high strangeness indeed!

The Weiner brothers and Charlie proved to be company of the highest calibre. Not only are they endearing people but extremely entertaining raconteurs. The twins tell tales of running off to New York City as fresh-faced teenagers and hanging out in the magical land that was Greenwich Village in the late 1960s. I also remember them talking of attending the famous Woodstock Festival in upstate New York, which ultimately heralded the end of the decade that saw the emergence of the inspiring new youth generation.

Jack, Jim and Charlie retold the events of the contact/abduction that they jointly experienced with their friend Chuck Rak, and it sent shivers down our spine; this was clearly something that still touched their lives on many levels and we could see why. Although it had clearly been a traumatic encounter, they remained strong and determined that it was not going to have a lasting, detrimental impact on their lives. I was truly inspired by how these men were dealing with their own experience after all these years and I came away feeling very different about my own. Additionally, it is worth stating that all three men have been fantastic artists since an early age and say that their contact experiences appear to have enhanced their creative abilities, whilst also seeming to develop previously untapped mathematical skills and scientific knowledge.

It was such a fantastic opportunity to get to know these people better and we spent the whole weekend doing just that. I had a relatively small support system at that time in the UK when it came to my ET contact, so talking with people who had either a direct experience of their own or an in-depth knowledge of the subject was so empowering and helped me to further develop my plan for moving forward.

I include this brief summary of our weekend in Connecticut for no other reason than to introduce you to some remarkable human beings. One of the reasons I finally decided to write about my contact experiences was for other people in my situation, and to show to them that one can survive and ultimately flourish from, something that had previously de-railed their life journey. Finding other individuals who can help that empowerment is vital. We are many but, more often than not, unaware of each other. There are people who would have us remain silent and unconnected, which only goes to prove that the contact that people like myself are experiencing is important on levels that we have not even begun to realise. I am not suggesting that we are more important than anybody else, although we seem to have been 'chosen' for this role of contact. We do therefore, need to create an unbreakable chain of communication and education with each other and the world we live in to see this responsibility through to its fruition.

*"Our lives begin to end
the day we become silent about the things that matter."*

Martin Luther King

Social Event: She Who is Called Manhattan

New York City – May 2004

In between our trips to Laughlin, Nevada in February and Newport, Rhode Island in August, we had another vacation to New York City in May. In a diary entry from that period I have found my description of our arrival that so reminds me of that moment that I hope you will allow me to share it with you:

> *"A rich and heady cocktail of voices – previously shaken, but now spiritually stirred – rose from the junction of Bleecker Street and MacDougal. Just like smoke caught in a slow swirling updraft it entered the open window of the apartment and circulated from room to room. It made me feel safe. It made me feel welcome. I was back in the arms of my ladylove – she who is called Manhattan. And, surprisingly enough, my wife was not jealous in the slightest."*

Even though the push and pull of trans-Atlantic travel was affecting our physical equilibrium and overall demeanor, the intense urge to walk the streets of Greenwich Village was just too much to resist. As we flowed into the steady stream of night-time people surfing the sidewalks that evening, a distinct feeling of 'coming home' touched us both. In a small bar on Bleecker we quaffed our first Margarita and listened to a live band playing a rock version of *My Favourite Things* from the *Sound of Music*.

"Only in New York City," I quipped over the top of the lead singer proclaiming allegiance – as the song says – to a varied list of fluffy items.

It was nearly a year to the day since Annie and I had first visited New York and our lives were now virtually unrecognisable to what they had been. A more informed view of our world, if not a million times more challenging, had replaced the relative innocence that we had known.

Over the past year I had written to and made friends with Debbie Kauble, formally Debbie Jordan of Budd Hopkin's book *Intruders*, where she went under the pseudonym of 'Kathie Davis'. Debbie is a successful writer, having co-authored *Abducted* with her sister in 1994, which is an extended description of her abduction experiences throughout her life. We exchanged emails weekly and spoke on the telephone numerous times and Debbie gracefully accepted our invitation to join us in New York for a week.

Debbie is a warm and friendly, colourful, larger-than-life character who has coped with the extenuating demands of her abduction experiences extremely well; even though one aspect of a particular encounter left her scarred in one eye, which requires medical attention to this day. She now lives in the mid-west in an area affectionately referred to as Hurricane Alley. After reading her book and listening to the story of her life, it was with great anticipation that Annie and I awaited her arrival.

When Budd had found out at the conference in February that Debbie would be joining us in New York he invited us both to present our separate stories at the Intruders Foundation meeting in May. Although a relatively small gathering of aficionados, to be invited to present at such a prestigious location was not to be missed and we both accepted.

During our first week in town, before Debbie arrived, Annie and I went to the cinema on several occasions, which is one of our major passions in life, not only when we are in New York. At one particular screening I was aware of a guy next to me who very obviously had a bout of flu, as he coughed and spluttered throughout the whole movie. I attempted to protect myself from his germ disbursement, but within a few days I too was to go down with a cold, which quickly turned into a debilitating dose of flu and would ultimately, although I was innocent to the fact until my return home, turn into pneumonia, which put me into a hospital bed for a week.

Therefore, on the evening of my presentation at the Intruders Foundation I was highly medicated but somehow made it through the proceedings. This was the first time that I began to talk about what I was becoming to believe the ET contact actually meant. Diversifying from the details of my abductions, I spent a good deal of time speculating on how we, the human race, should really be responding to this event by turning our contact experience into an opportunity to review our lives and our place on this planet. I began to explore the possibility that the ETs were 'holding up a mirror to us' so we could reflect upon how we were conducting our lives. I was absolutely amazed at how well the audience reacted to the simplicity of my words. They didn't want to hear about aliens, UFOs and abductions; they were responding to the deeper, more philosophical question of why and, even more productively, how can we respond to it.

I was truly amazed and overwhelmed by the love that came back from these people, the questions that were raised and the apparent insight and awareness they were experiencing of their own circumstances. I believed on that night that I was finally on the right path that would help me to eventually discover and understand why it was important for me to talk about my experiences.

Even though on that night I was feeling like death warmed up and felt absolutely dreadful on my journey back to England due to my undiagnosed pneumonia – an occurrence that I would not want to repeat – my heart was filled with such joy for I realised that there was actually hope for the human race to escape from their self-inflicted chains, the same chains that had hampered and stilted our emotional and spiritual growth for so very long.

"Fears are educated into us,
and can, if we wish, be educated out."

Dr. Karl Menninger

Experience: Fear – No Longer an Option

Part 1: The Torn T-Shirt

Essex, UK – Early June 2004

This particular mid-week evening was no different from any other: Annie had gone out to visit friends, Reece, my son, was in the study working on a homework project for school and I was in the lounge watching television. I knew that my son's work was important and had to be given in the next day so I periodically popped in on him to ensure he remained focused and did not get distracted by video games. Sometime during the middle of the evening I began to feel weary, so I retired to my bedroom to watch TV. On my way, I looked in on Reece one final time and told him what I was planning to do. He informed me that all was well and that he would definitely stay up to whatever time necessary to ensure completion of the work.

I turned on the television and laid on the bed. I went back to the film that I'd been watching downstairs. It was just before 9.30 p.m. I don't actually remember feeling tired enough to sleep but within a few minutes I began to feel drowsy and was unable to focus on the TV anymore. I began to experience a most peculiar sensation; a repetitive internal buzzing started, which began in what felt like the inside of my head and pulsed out to my forehead and the back of my neck. It came in rhythmic bursts of three, repeating itself every minute or so. I knew this had happened before. I then felt what can only be described as an electrical pulse that began in my shoulders, tensing and flexing the muscles of my arms, twitching all the way down my body and finally leaving via my feet in a spasmodic jolt. It continued in a predictable pattern of waves that had my body gently convulsing uncontrollably. I recognised these sensations as an opening gambit on the part of the ETs to making contact with me. I closed my eyes in tense anticipation of what might possibly follow. There was a palpable alteration to the air pressure in the room; which made me unintentionally and momentarily catch my breath and swallow hard. In doing so I became aware of a smell, one that I had experienced before on these occasions, one similar in odour to the smell given off by an over-worked photocopier machine.

The internal buzzing intensified and I had the illogical impression that my body was beginning to somehow fold in on itself and collapse. At this point I must have lost consciousness because when I next became aware of my surroundings it was apparent that I was no longer lying in my bedroom. It was clear that the predicted transportation of my physical body had taken place and I was once more a guest of my alien hosts.

I had a sense of compounded anger reflective of my growing desire to stop feeling like a victim of these seemingly forced circumstances, and to somehow become an active and vocal participator. My contact had changed over the years: in the beginning I simply accepted what I experienced with no preconceived judgment because my belief systems were unadulterated and hadn't yet been polluted. As such I didn't respond with fear. As time went by though and my perception of these occurences became corrupted by the influence of the modern world, I started to experience them as fearful events, although in fact the reality of my contact, upon reflection, remained the same; it was only my perception of them that had changed. Additionally I was dealing with a need, at this later date, to sometimes departmentalize certain experiences as they happened, therefore I would forcibly forget certain events and then deal with the nagging influence of something that is experienced but then hidden away.

I was lying down. As usual my eyes were firmly, and consciously, closed shut. It was always my intention to avoid seeing or interacting with my situation for as long as possible – keep your eyes closed and it won't exist! It felt as if I might be naked because I could feel the coolness of whatever I was lying on pressing against my back and legs. All around me I could hear activity, as if a number of people were scurrying to and fro. I felt scared. Intellectually I should have known that there was no need for me to be fearful; after all, I had never, ever come to any harm during one of these events. But my response was primeval, not rational.

I experienced the sensation of my body being drawn down, being pulled by some invisible force, as if I was becoming part of what I was lying upon. It felt as if I were melting away. Then, just as quickly, I began to feel solid again, slowly becoming cohesive and experiencing structure. I became aware of my environment, breathing the air, feeling the weight of my body. I understood who I was. I knew why I was there. And then, in an instant, I slept; a deep, comforting human sleep enfolded me and I sank deeper and deeper.

My son's memory of the evening goes something like this: he wasn't looking forward to working on the homework project because, as usual, being the teenager that he was, he had left it to the night before it was supposed to be handed in to start working on it. He didn't mind the fact that I kept looking in on him, but he felt more relaxed when I announced that I was going to bed and that I would see him in the morning. After another hour or so he started to feel really drowsy and found it hard to stay focused on the job at hand. He knew how important it was that he finished what he was doing, so he got up and walked about to wake himself up. After going downstairs and getting a drink, Reece looked in on me to let me know what he was doing, but I wasn't in my bedroom where he had expected to find me. Thinking that I was probably in the bathroom he returned to the study. Within minutes, however, he felt so exhausted that he had to lie down on his bed for a few minutes to refresh himself. Looking back now he appreciates that it was a ridiculous action on his part considering the importance of completing the project but the urge to do so was too strong and too irresistible – my son would later comment that he felt he had been 'forced' to go to sleep. The next thing Reece remembers is Annie coming home later on that evening.

To complete the picture, this is Annie's memory of the evening: just before seven o'clock she had gone out to visit friends, with Reece doing his homework upstairs and me stretched out on the couch watching TV. She returned home just after midnight and the first thing she noticed as she parked the car, something that struck her as quite odd, was that our two bedroom windows were wide open. It wasn't an extremely cold evening, but the fact that the windows were open so wide seemed puzzling to her. Entering the house she noticed that a lot of house lights were still on, an unusual sight considering the lateness of the hour and her belief that both Reece and I were probably asleep. She came straight upstairs because, in her words, 'something felt strange'. As she walked into our bedroom she was confronted by an extremely disturbing image: there I lay on the bed, on top of the covers, wearing an unidentifiable – but obviously female – torn t-shirt and a pair of underpants. A cold breeze was blowing in from the gaping windows and the main light and the side lamps were blazing away. I appeared to be fast asleep, but the most alarming aspect of this surreal picture was that blood was oozing from my nose and onto my moustache and beard and had dribbled all over the front of the unrecognizable garment I was now wearing. Shaking me awake with some difficultly she quickly established that despite the relatively profuse amount of blood, I appeared well enough to be roused and communicate with her. It was, she reassuringly convinced herself, just a very bad nosebleed (which I had experienced on previous occasions). In order to shed a bit more light on the matter Annie went into Reece's bedroom to find him fully clothed and fast asleep on top of his bed. Waking him up she asked if he knew what had happened to his dad. Reece got to his feet, still groggy from his apparently enforced sleep and followed Annie back into our bedroom. Annie said that Reece just stood there gaping at me, apparently trying to take in the bizarre image that confronted them both. Clearly Reece was initially disturbed by the blood he saw on my face, but once he too had established that the situation was not as bad as it appeared he came out with a classic line that will forever bring humour to the memory of what will remain an out-of-this-world event.

Looking down at me, lying prostrate on the bed, blood all over my face, blurry-eyed and confused to boot, he looked me up and down, smiled cheekily, cocked his head to one side and said: "Dad, what on Earth are you wearing?!" Only a teenage boy could have said it. In fact, knowing my son, only *he* could have said it.

Annie used a pack of bathroom wet-wipes to clean up the blood on my face and placed them in a pile on the side in the bathroom. It was her intention to pursue some sort of evaluation of the 'blood' as it did appear to have a peculiar pink hue to it and a strange unfamiliar aroma. How she would go about this, she had no idea, but she did consider contacting one or two of our UFO friends for advice.

After getting Reece back to bed and making sure that I was now cleaned up and in good health, Annie prepared the house for the night and came to bed herself. When she went back in to the bathroom in the morning, the pile of used wet-wipes had gone. Now, you might suggest that one of us got up in the night, stumbled into the bathroom and absent-mindedly flushed them away; a reasonable suggestion except that Annie and I always use our en-suite bathroom at night thereby eliminating us, and Reece is a deep sleeper and once asleep does not normally stir again until the sun comes up. I'm not necessarily suggesting something paranormal happened here or that the ETs 'came back for the evidence'; all I'm saying is that they disappeared.

As for the woman's t-shirt I had been returned in, I still have it, still in its ripped state and still bloodstained. We have since heard from other sources that this is not the first time an Experiencer has been returned wearing different clothes to the ones they were taken in. I don't know if there is some sort of communal area designated for the undressing and redressing of their guests, but I really do feel that the ETs should begin issuing a ticket to the Experiencers as they are de-robed, just like a hat-check girl in a club!

I have no clear recall of what played out once I had been taken, except for what I have already said. The one lasting sense of the event was my increasing need for things to be different; I wanted to stop feeling like a victim. It was important that if this contact was to remain a part of my life I had to somehow start having a say in what happens to me.

"Be not afraid of life.
Believe that life is worth living,
and your belief will help create the fact."

William James (Philosopher and psychologist)

Experience: Fear – No Longer An Option!

Part 2: Breakthrough

Essex, UK – Late June 2004

My first conscious memory of this particular contact experience was waking up and being there. Annie and I had gone to bed that evening, we both read for a while and then switched off the light. There wasn't anything to suggest that I would be 'visited', no tell-tale signs manifesting themselves, no hint or suggestion that I wouldn't have a reasonably good night's sleep and wake up the next morning as normal.

I woke up and knew instinctively that I hadn't been to sleep for very long. I also knew – in a heartbeat – that I was not lying in the warm and cosy confines of my own bed. I was naked, uncovered and lying on top of what appeared to be a hard, unyielding, flat surface. There was an instant familiarity with my surroundings: I had been taken. Turning my head just a few degrees to the left and right of where I lay, I could see the recognizable smooth texture of the walls and the domed ceiling above my head. Whatever was lighting the room gave everything a peculiar soft fuzzy edge and hindered the perception of depth and distance. One thing was clear though; I was not alone.

Even before I tried to focus on the movement I could tell that there were several figures moving almost imperceptibly all around me. From past experiences I knew that the table on which I was lying was no more than a couple of feet off the floor and I could see that my companions were a bit taller than that. Around the edge of the room appeared to be several freestanding units that reminded me of bedside tables and which were of particular interest to several of the figures. They moved from one to the other with apparent precision and without bumping into each another. Bending over the surface of each unit, an action was carried out after which they proceeded to another in a beautifully choreographed flow of movement. Their actions were so fast that I was unable to determine how many of these little figures were present in this activity, but estimated a good dozen of these fellows whizzing about. I felt like a detached observer and it seemed as if I was invisible to all present. That was until I decided to move.

At times like this I often feel as if I have been anesthetized, as if the movement of my body is being restricted by an unknown agent. This isn't always the case, but on some occasions a feeling of drowsiness is experienced that normally impacts upon my thinking or movement. At some point in the procedure though, as if something is administered to refresh and animate one's clarity of thought or movement, a Grey will either place one of their long-fingered hands upon your head or move their face close to yours and stare with their large, black, hypnotic, eyes and you are somehow released. But on this occasion I felt wide awake and in total control of my mind and body. My senses felt sharp and energized and I imagined that if I had wanted to I could have sprung from the table with no assistance whatsoever.

Sliding my elbows back along the side of my body, I bent my arms and pushed my upper torso off the bed until I was sitting up. In the blink of an eye, all movement in the room stopped. As if I had been watching a movie and had pressed the pause button the picture around me instantly froze. I swivelled my eyes from left to right and craned my head behind me to see that each and every figure in the room was caught in mid-movement. What I found most intriguing of all was that they were all staring at me. Whatever they had been doing, they had stopped and their faces were all turned so as to look directly at me.

I could feel that I was smiling; it just made me want to chuckle. I had caught them out. My movement was apparently the last thing they had expected to happen. Although these little guys never really show any emotion or feeling, I could sense what was going on for them. It was as if an energy hung in the air charging each molecule with a jolt of electricity. I suppose the nearest and best translation of what I sensed coming from them was "What the f**k!"

Feeling so empowered with the situation, unlike the normal experience of being controlled, I found I had a voice. "Yes guys, I can move." I waited. "You didn't expect that, did you?" Another spark of energy, proving once more that my new outburst was only adding to their sense of bewilderment. I lay there propped up on my elbows, wondering if I should move further, waiting for some sort of signal on how to proceed; after all this was unprecedented.

And then, from some unknown source, an invisible instruction was apparently sent out, for the collected group broke their pose and once more moved again in unison. Several of them shuffled towards me with obvious hesitancy. They stopped within about two feet of where I was and stared. I could see the remaining figures filing out of the room like a shiny grey crocodile into the darkness of the supposed corridor beyond. I sensed that I was being asked to lie back down, but instead – just for the hell-of-it actually and wondering how far I could take this thing – I swung my legs to the right until they hung down skimming the floor. They stepped back. I found my voice softly resonating in the small domed room: "I'm not afraid anymore guys. I'm not scared of you."

From the aperture in the wall appeared another figure. Taller than the rest and certainly not showing any sign of hesitation it moved directly towards me and stopped within just a couple of inches from my face. Probing deeply into my unblinking eyes, which stared back with determined conviction, I heard her voice loud and clear. It was The Witch, her head at exactly the same level as mine as I sat squarely opposite her.

"What has happened since last we met?"

I didn't answer, but just looked back.

"What has happened to you? What do want from us? How can we

help you?"

As if some sort of floodgate had been opened, her questions rushed my senses and overwhelmed me as they probed every nook and cranny of my mind seeking clarification and explanation of my unexpected actions. Her companions stepped closer and stood by her side. I struggled initially to understand what I was sensing; what was being exchanged between them, until I realised that the best way to describe it was *excitement*, as if my behaviour had created a sense of wonderment within them.

Once again, the questions – reshaped, reworded, restructured – came flooding into me: What had happened to me? How was I different? What was my intention? What did I want to happen now?

In return I just smiled once more.

What? Why? How? Explain? Understand?

"I'm not afraid anymore. I want to be a part of this. I have to have respect," I whispered.

"There has never been a need for you to be afraid; that was your choice. You have always been integral to everything we have done. At all times there is love."

The Witch's hands came out and helped me back until I was once more lying flat down. Instantly I was surrounded by a group of her assistants. More excitement. Great enthusiasm. Instructions. Responses. Hands were moving up and down my body, over my face, fingers tracing and pressure applied. Again I heard her voice, but there were no words, just an understanding of what was now occurring. I had been brought there for a reason, but that was now redundant because of what I had done; something else was now going to happen but they needed my permission first before they could proceed. She made me feel it was all right, that it was needed and that I would be happier if it happened. I would be improved.

Without understanding, but knowing in my heart that I was safe, I consented. My memory of the procedure that was then carried out is hazy, as I believe part of it might have been either painful or somehow unpleasant, but I had agreed to it and I am still here in one piece.

When I was finally returned to my home and I had time to process what had happened, my sense of it was this: I believe that

I was somehow altered; that something – possibly an implant – was removed and another one was put in its place. An upgrade if you will, because it was from that time that I changed in some very basic ways regarding what I could no longer eat or drink (no red meat, processed food, non-organic food or alcohol). Additionally, my psychic abilities became heightened and much more refined, showing themselves especially in my interactions with other people and my ability to read and assess their thoughts and feelings. My understanding of how to proceed came into sharp focus and I was able to form a sort of game plan about how to communicate my feelings with regard to my ET contact. I began to understand that contact is about reminding us who we are and what we must now do. And...most importantly, I was definitely no longer afraid.

"The whole course of human history may depend on a change of heart in one solitary and even humble individual – for it is in the solitary mind and soul of the individual that the battle between 'good' and 'evil' is waged and ultimately 'won' or 'lost."

M. Scott Peck

Food For Thought: Memory

Part 3: Non-Observable Spatial Dimension

The inevitable question is that if memory is not inside the brain, where does memory reside? The short answer is: officially we don't know. Scientific pursuit has always been looking for evidence to support a logical conclusion derived from a general theory. If the general theory is fundamentally flawed, the progressing of science will stop and wait for convincing evidence to overturn the general theory. Only from there will science flourish again on the new foundation. Recent research in quantum mechanics suggests that *reality*, in a more basic level, may not be material at all. Memory may be in a physical form we do not understand, or it may even reside outside our physical world. We shall wait for new evidence to emerge. And that evidence might be closer than we imagine – it may just be right there, in the forest, amongst the trees.

Could it be that our memories actually dwell in a space outside our physical structure? Biologist, author, and investigator Dr. Rupert Sheldrake notes that the search for the mind has gone in two opposite directions. While a majority of scientists have been searching inside the skull, he looks outside. According to Sheldrake, author of numerous scientific books and articles, memory does not reside in any geographic region of the cerebrum, but instead in a kind of field surrounding and permeating the brain. Meanwhile, the brain itself acts as a decoder for the flux of information produced by the interaction of people with their environment. In his paper 'Mind, Memory and Archetype Morphic Resonance and the Collective Unconscious' published in 1997 in the *Psychological Perspectives Journal*, Sheldrake likens the brain to a TV set, drawing an analogy to explain how the mind and brain interact, he states: 'If I damaged your TV set so that you were unable to receive certain channels, or if I made the TV set aphasic by destroying the part of it concerned with the production of sound so that you could still get the pictures but could not get the sound, this would not prove that the sound or the pictures were stored inside the TV set. It would merely show that I had affected the tuning system so you could not pick up the correct signal any longer. No more does memory loss due to brain damage prove that memory is stored inside the brain. In fact, most memory loss is temporary: amnesia following concussion, for example, is often temporary. This recovery of memory is very difficult to explain in terms of conventional theories: if the memories have been destroyed because the memory tissue has been destroyed, they ought not to come back again; yet they often do.'

Sheldrake goes on to further refute the notion of memory being contained within the brain, referring to key experiments that he believes have been misinterpreted. These experiments have patients vividly recalling scenes of their past when areas of their cerebrum were electrically stimulated. While these researchers concluded that the stimulated areas must logically correspond to the memory generated and recalled, Sheldrake offers a different view: '... if I stimulated the tuning circuit of your TV set and it jumped onto another channel, this wouldn't prove the information was stored inside the tuning circuit,' he writes.

So if memory does not live in the brain, where does it reside? Following the notions of previous biologists, Sheldrake believes that all organisms belong to their own brand of form-resonance – a field existing not only inside but also around an organism, which gives it instruction and shape. Sheldrake considers the morphogenetic, or form-shaping approach to be an 'alternative to the mechanist/ reductionist' understanding of biology, which is the predominant view. The morphogenetic approach sees organisms intimately connected to their corresponding fields, aligning themselves with the cumulative memory that the species as a whole has experienced in the past. Yet these fields become ever more specific, forming fields within fields, with each mind – even each organ – having its own self resonance and unique history, stabilizing the organism by drawing from past experience. 'The key concept of morphic resonance is that similar things influence similar things across both space and time,' writes Sheldrake.

The holographic theory as theorized by Karl Pribram states that memory resides not in a specific region of the cerebrum but instead in the brain as a whole. In other words, like a holographic image a memory is stored as an 'interference pattern throughout the brain'. However, neurologists have discovered that the brain is not a static entity, but a dynamic synaptic mass in constant flux – all of the chemical and cellular substances interact and change position in a constant way. Unlike a computer disc which has a regular, unchanging format that will predictably pull up the same information recorded even years before, it is difficult to maintain that a memory could be housed and retrieved in the constantly changing cerebrum. But conditioned as we are to believe that all thought is contained within our heads, the idea that memory could be influenced from outside our brains appears at first to be somewhat confusing. On his *Staring Experiments* website, Sheldrake writes that: '… as you read this page, light rays pass from the page to your eyes, forming an inverted image on the retina. This image is detected by light-sensitive cells, causing nerve impulses to pass up the optic nerves, leading to complex electrochemical patterns of activity in the brain. All this has been investigated in detail by the techniques of neurophysiology. But now comes the mystery. You somehow become aware of the image of the page. You experience it outside you, in front of your face. But from a conventional scientific point of view, this experience is illusory. In reality, the image is supposed to be inside you, together with the rest of your mental activity.' (www.sheldrake.org/experiments/staring)

Yet another area for consideration for the relatively local storage location of memory outside the physical form is the Human aura/auric field. Much has been written about the energy of living things but up until recently there has been little scientific evidence gathered.

The energy field surrounding living things has proved difficult or impossible to measure using past scientific techniques. However, science and spirituality are on a convergent course. Eventually, we will have instruments, if we don't have already, that can reflect an individual's state of balance. When we turn the corner from science we must consider the universal characteristics of energy. Quantum physics states that energy and matter are interchangeable with the String Theory suggesting that differences in physical matter are simply variations in energy vibrations. In a similar fashion, each human is composed of the divine energy of the Soul in the form of body, thought, and spirit. Energy does not emanate or reflect from a person, the energy is the person, the core. This understanding is fundamental to maintaining your energy field and body in harmony. Since the body is a manifestation of human energy, dis-harmony in the energy field will cause disease in the body. If the human energy field is out of balance, the body will be out of balance.

While the search for memory challenges traditional biological understanding, investigators like Sheldrake believe that the true residence of memory is to be found in a non-observable spatial dimension. This idea aligns with more primal notions of thought such as Jung's 'collective unconscious' or Taoist thinking that sees the human mind and spirit derived from various sources both inside and outside the body, including the energetic influences of several different organs. In the Taoist view, the brain does not act as a storage facility, or even the mind itself, but as the physical network necessary to relate the individual with its morphic field, an expression used by Sheldrake in his 1981 book *A New Science*, where he uses the term to refer to what he thinks is 'the basis of memory in nature...the idea of telepathy-type interconnections between organisms and of collective memories within species'.

If our memory is not to be found ensconced within the physical structure of the brain, and that somehow the record of our memories is stored in some sort of electrical field, either one 'surrounding and permeating the brain' as Sheldrake suggests or in the electromagnetic Aura/Auric Field that surrounds our body, then maybe it would be worth considering that there is a further, much greater, body of record that not only holds the complete memory of the individual, but also the record for every single human being alive or dead.

In many cultures around the world such a thing is referred to as the *Akashic Records* (a term commonly used in Theosophy and Anthroposophy). These records are reputed to contain all knowledge of human experience and the history of the cosmos. They are metaphorically described as a library and other analogies commonly found in discourse on the subject include a 'universal computer' and the 'Mind of God'. Descriptions of the *Akashic Records* assert that they are constantly updated and that they can be accessed through astral projection. The concept of the *Akashic Records* originated in the Theosophical movements of the 19th century borrowing heavily from Hindu tradition and philosophy and remains prevalent in New Age discourse. The records are of all that has happened and are impressed on a subtle substance called Akasha (or somniferous/sleep-inducing ether). In Hindu mysticism the Akasha is thought to be the primary principle of nature from which the other four natural principles, fire, air, earth, and water, are created. These five principles also represent the five senses of the human being.

According to believers, the Akasha is the library of all events and responses concerning consciousness in all realities. Every life form therefore contributes and has access to the Akashic records. Any human can become the physical medium for accessing the records, and that various techniques and spiritual disciplines (e.g., yogic, pranayama, meditation, prayer, visualization) can be employed to achieve the focused state necessary to access the records. Just as conventional specialty libraries exist (e.g., medical, law), adherents describe the existence of various Akashic records (e.g., human, animal, plant, mineral, etc.) that in their summation encompass all possible knowledge. Most writings refer to the Akashic records in the area of human experience but adherents believe that all phenomenal experience as well as transcendental knowledge is encoded therein.

The most famous example of someone whom many claim could successfully read the Akashic records is the late American mystic Edgar Cayce, who was able to access these records whilst in a sleep-like state or trance. Dr. Wesley H. Ketchum, who for several years used Cayce as an adjunct for his medical practice, described Cayce's method: 'Cayce's subconscious is in direct communication with all other subconscious minds, and is capable of interpreting through his objective mind and imparting impressions received to other objective minds, gathering in this way all knowledge possessed by endless millions of other subconscious minds.' In the book *The Law of One: Book One*, a channelled entity, identifying itself as Ra, stated in 1981 that Cayce himself did, in fact, channel the Akashic records, as opposed to an entity. Other individuals who claimed to have consciously used the Akashic Records include: Linda Howe, Charles Webster Leadbeater, Annie Besant, Alice Bailey, Samael Aun Weor, William Lilly, Manly P. Hall, Lilian Treemont, Dion Fortune, George Hunt Williamson, Rudolf Steiner, Max Heindel, Madam Helena Petrovna Blavatsky, Edgar Cayce, William Bhulman and Michael J. Dickens.

Although not a fundamental part of my contact experience, where memory resides and how it is accessed is an integral and relevant section within the story. Because my memory of certain experiences, and elements within others, has either been suppressed or manipulated at times – for my self protection – it does make it important for us to investigate the bigger picture. Based on my communications with certain ETs, I believe that to allow full or partial memory of all contact events, all of the time, would have been dangerous for the individual on many levels. After all, we are here to live our life, our human life. However, having said that, I believe that we are now on the cusp of fully ascending to a higher level of spiritual and intellectual understanding and appreciation. Accordingly, we can now embrace the concept and reality of being in contact with an advanced intelligence. I don't believe, that before this moment we could have dealt with the 'distraction'. The idea, that has been 'doing the rounds' for many years of the human race being unable to cope with the reality of ETs because it would destroy many vital parts of our human belief and working infrastructure (such as religion, history, science, politics, economy) is not true and has been propagated by forces unknown as a continued way of controlling and suppressing certain elements of human evolution. Please understand that I am translating a much greater and larger truth here, so it can be read and understood within the confines of a few paragraphs. I hope this section is read in the spirit in which it has been written. If at times I appear to sound patronizing please believe me that that is not my intention, merely arising from my desire to be more 'user-friendly'.

Through my research and my intuitive skills of deduction and logic, I am happy to settle on the concept of an electrical energy field that exists, either around the brain exclusively or around the entire body, where memory resides and that certain parts of the brain are responsible for translating and processing specific memory types – sensory, short term and long term. I have put my new appreciation of this subject to our ET visitors and they have not contradicted me, so I guess that, if not necessarily right in the bull's eye, I might be at least hitting the target. After due consideration I truly believe that I have started to ask the right questions at last.

"Star light, star bright, first star I see tonight.
I wish I may, I wish I might,
Have the wish I wish tonight."

Traditional nursery rhyme

Experience: Our First Trip To Anne's Home

Newport, Rhode Island – July/August 2004

Part I: Little Star

After meeting Anne Cuvelier at the UFO Congress Conference in Laughlin, Nevada early in the year, Annie and I were absolutely over the moon to receive an invitation from her to join the annual gathering of UFO aficionados at her home in July. Anne owns an architecturally stunning home with black and white timbers and pale blue cladding, that sits right on the water's edge overlooking the sailing boats of Newport harbour. It was built in 1870 and has remained in the family since that time. The beautifully tended gardens are complemented by a salt-water swimming pool and hot tub, with an ornate pier that stretches down to its own boat jetty; complete with Anne's little speed boat tied up there. The house is just a five-minute walk from the centre of Newport, which in itself is quite the most exquisite example of New England architecture and history. I shouldn't really tell you this as I would prefer to keep this magical haven a secret, but anybody can stay there at any time of the year because it is actually a luxurious bed and breakfast establishment. (www.sanford-covell.com)

There is a central core of guests who attend every year, many of whom we had already met and become good friends: Budd Hopkins

(UFO and Abduction Investigator and author), Dr John Mack (Harvard professor, abduction investigator and author), Jack and Jim Wiener, along with Charlie Foltz (Experiencers) – all central characters from the book *The Allagash Abductions* (written by Raymond E Fowler), Linda Cortile (Experiencer), Donald Matrix (Investigator and Experiencer) Dr. Michael O'Connell (Investigator, Hypnotherapist, Harvard graduate and clinical psychologist), Barbara Lamb (Abduction investigator, hypnotherapist, Crop Circle expert and author), Paola Harris (UFO and abduction investigator and author) and Bob and Terri Brown (directors of the UFO Congress Conference) to name but a few. Anne tries to include new guests each year to ensure an influx of new blood and after I had spoken at the 'Experiencer's Open Mic' presentation at Laughlin that year, Anne thought that my inclusion in her guest list that year would add some additional colour to the proceedings.

Before driving up to Rhode Island, Annie and I went to stay for a few days with our dear friends John and June Franklyn who were regular participants at Newport each year. At the time they lived in upstate New York in a beautiful detached home tucked away in the woods surrounded by roaming deer and humming birds outside and a pack of tiny, yet masterful, Yorkshire Terriers inside. John is a man of rare gentleness who stands well over six feet tall but would appear that way even without his extended height. Although now retired, at the time he was a successful high school teacher of long standing and the type of educator that we all wish we could have had during our younger years. His wife, June, is an 'all-American' wife and mother, who simply oozes love and care for everybody she comes into contact with. After spending time with her, you cannot help but come away feeling cherished. Both of them are rare individuals and we are blessed to call them our friends.

We arrived at Anne's after lunch on Monday, unloaded our bags and were shown to our accommodation. Unpacking our bags I explained to Annie that I felt a little apprehensive about meeting everybody. She wondered why that was, considering that I already knew most of the guests. I wasn't sure why, but there was something niggling away at the back of my mind suggesting that I should be hesitant, although I had no idea why or of what. I would soon find out the source of that intuitive uneasiness.

Standing up on the elaborate wooden porch at the rear of the house surrounded by luxurious wicker sun loungers and small glass-topped tables, I looked down onto the garden area and to a closely huddled group of people clad in swimming costumes, laughing and gesticulating, whilst they dangled their feet into the steaming hot tub as it hubbled and bubbled. From that distance and in the bright sunshine I squinted my eyes so as to identify the sunbathers. Out of the eight people who sat there I recognised most of them except one particular man and woman. Anne had told me of an Irish couple, who now worked and resided in Canada, who would be attending: a man from Galway on the west coast of Ireland called Sean, an architect, and his wife Sapphire who originally came from Cork, also on the west coast, who was a teacher. This must be them, I thought. There was something about the woman that seemed familiar, but I wasn't immediately sure what that was. She sat there in her swimming costume, tall glass of iced drink in her hand and a large floppy sun hat pulled down over her eyes. Her manner was very colourfully animated and it wasn't long before she placed her drink to one side as she chortled excitedly and clapped her hands together in a very familiar fashion; a style in which I had witnessed many times before. With her face virtually covered by the brim and shadow of her hat I strained to see her features properly and to connect with the thoughts that were starting to race through my excited mind. I stared harder as this slim, dark-haired young woman squealed in delight at another moment of intense humour exchanged within the group.

It couldn't be.

Yet again, she raised her two hands up in front of her face in a praying position and breathlessly clapped them together like a cute little seal waiting to be thrown a fish.

Could it be her? Could it?

Annie and I climbed down the stairs to the garden and walked over to where everybody was sitting in the sun. As we approached them the young woman threw her head back and sent a wave of laughter upwards towards the clear blue sky. I saw her face clearly. In that instant I knew; not one hundred per cent, because the truth was just too surreal to take on board, but near enough; near enough to be able to say in my mind: *'It's Little Star.'*

We were introduced to those we didn't properly know and smiled and greeted those we did. Sapphire looked directly at me and immediately dropped her face back down so she was hidden by the brim of her hat. I thought there was a moment of recognition; only a split second, less than that really, but there was definitely something there. Or was I kidding myself? Were just the physical mannerisms of this woman persuading me of something that was not fundamentally true? Had I seen Little Star at this age yet? I wasn't sure. In the intensity of the moment I couldn't remember. Long dark hair, slim body, excited laughter, animated behaviour, ... the clapping? There was something more though; something much deeper, something much more profound, some intrinsically intuitive and innate. Something that I could not put into words.

Annie and I retreated back to our room and changed into our swimming costumes. Upon our return to the pool area Sapphire and Sean had gone to have a siesta. We spent the next couple of hours swimming in the pool and getting to know everybody else better. At one point I sat in intense conversation with a lady from Vermont called Eva. She too was an Experiencer and had worked closely with Dr John Mack exploring what had happened to her. We shared details of our experiences. I told her all about my contact with Little Star: the petite child I first encountered; the singing of the song 'Twinkle, Twinkle, Little Star'; the animated clapping; the levitating building bricks; calling one another Little Star and Peter Pan. Eva listened intently, not interrupting me and taking in every single detail. When I had finished she smiled warmly with a knowing look on her face. "What is it, I enquired?" She looked down as if she was thinking of what to say next with great consideration. I realised the importance of giving my new friend the opportunity to process her thoughts without my interruption, so I waited.

Finally, she looked up and said, "I know who Little Star is".

With what had already happened earlier, after seeing Sapphire by the pool, this was starting to feel like quite a surreal episode rolling out before me. I wasn't sure if I really wanted to hear what Eva had to say; well, I did and I didn't. My mind was spinning in turmoil. Eva waited patiently for a response; she clearly was not going to say anything without my permission. Finally, with obvious trepidation, I said, "Okay... go ahead." Eva raised her eyebrows and fixed me with a look that seemed to say, 'Here it comes.'

"It's Sapphire."

"How do you know, I asked?"

"I just know."

My jaw was on the floor. I know I had already boarded that train of thought and clearly I had reason to believe it by what my eyes were telling me. How on Earth could Eva have possibly come to that conclusion? The answer to that question lay somewhere in the future; for as my intuitive abilities became more honed I was to become aware that I too had the facility to tune-in to the experiences of others, and not just those involved in the world of ET contact.

At dinner I kept catching Sapphire's eye, but she would timidly drop her gaze to the floor and look away. Why couldn't I categorically state that this was Little Star? I now remembered that I hadn't seen her for quite a while, maybe even since her late teens in fact. Also, when you see somebody you know outside of his or her regular environment, somebody from work in the supermarket or something like that, it throws you a bit. Ultimately though, I needed to appreciate that when I had been with Little Star it was always in a surreal environment and state of mind and my recall of specific events we were involved in was never like remembering something ordinary, not like recalling a conversation you had at work nor a programme you saw on television nor playing with your children at home or in the park, this was something completely beyond the norm. So, I needed to go a bit easier on myself.

As people started to retire after dinner, Sapphire and Sean stood by the door leading from the porch to the house and said their goodnights. Suddenly, Eva called out to Sapphire, "What was your favourite nursery rhyme when you were small?"

Sapphire looked puzzled. "What do you mean?" she replied.

"Was it 'Twinkle, Twinkle Little Star' by any chance?"

Sapphire turned abruptly and pulled her husband through the swinging door and rapidly disappeared. People looked from one to another with quizzical looks, but Eva just sat there and smiled. She gestured to me to follow her around to the side of the porch. In a whispered tone she said, "I spoke with Sapphire earlier on and guess what?"

"I don't know," I replied hesitantly.

"Well, she won't admit that she totally recognizes you but after some gentle and subtle probing questions she told me that when she was very young she had an invisible brother called…"

The silence was pregnant with our joint anticipation.

"Peter?" I whispered.

Eva nodded. "She was so infatuated by the name Peter," she went on somewhat smugly "that she called the family dog Peter – *Shadow Peter* – whilst everybody else called it by another name."

I didn't say anything.

"She also said that her favourite toy when she was little was her different-coloured building bricks and that they are the only keepsake she still has from those days and they're sitting wrapped up in the loft at home."

How could I reply to this? So I still said nothing in return.

Eva carried on. "Sapphire said that when she first saw you standing on the porch watching her sitting by the hot tub she just had the overwhelming sensation to jump up and run about like a little child, clapping her hands and whooping uncontrollably."

I nodded. I understood.

"What you don't know Steven is this: a couple of years ago a few of us, including myself, were abducted on the final night of one of Anne's little gatherings and what transpired really shook Sapphire up. She wasn't here last year, but this time around she has made it very clear to me that she needs to keep a very low profile, that she doesn't want to 'conjure' anything up, that she desperately needs to have a nice quiet week just relaxing and talking with old friends."

I tried to think about what this all meant. This was something beyond my wildest dreams; here I was finally face to face with Little Star. Here on Earth, away from the ETs, we had finally been drawn together. What did it all mean? Clearly this was meant to be, surely?

Lying in bed that night listening to the varied sounds of boats in the harbour, cars making their way across the suspension bridge to Rhode Island and the water just a little way from where we were sleeping, I talked the matter over with Annie. We came to the conclusion that we would have to tread delicately as we didn't want to alarm Sapphire... Little Star... and scare her away. We needed to wait to see how she wanted things to progress, if she wanted to at all.

When I worked with Budd Hopkins the previous year at his home in New York City, we had recorded the regressional hypnosis session that he conducted where we had specifically covered, along with the David Bowie Night Abduction, the very first encounter I ever had with Little Star. I had brought the tape along with me and Anne, our wonderful host, had asked if she could listen to the recording. We set everything up in her private study so Anne, Annie and myself could listen. What I didn't know was that we were going to have an additional guest.

As we were about to start Sapphire arrived. "Is this all right with you Steven? Can I sit in with you guys and listen to your regression?"

A verbal response somehow eluded me. I smiled like a simpleton and nodded an agreement.

We played the tape.

About half way through the recording I became very emotional. I had to leave the room. I made my way back to our bedroom and sat on the edge of the bed with my head in my hands. This was all becoming too much. I felt completely overwhelmed. And then I realised I wasn't alone. I looked up to see Sapphire standing there. I looked at her through eyes filled with warm sweet tears. I stood up and she moved towards me. Taking tentative steps, actually and metaphorically, we stepped into each other's arms and hugged tightly.

She whispered softly and familiarly, "It's all right Peter, I'm here, your Little Star is here now."

"Millions of spiritual creatures walk the Earth unseen,
both when we wake and when we sleep."

John Milton (*Paradise Lost*)

Experience: Our First Trip To Anne's Home

Newport, Rhode Island – July/August 2004

Part 2: The Last Night

The rest of the week was such a joy. The time was spent hanging out with our old and new chums; philosophising about the mystery of the contact experience and sharing stories about our encounters and generally being astounded by the experiences of others. This was the only year that we were to spend time at Newport with Dr John Mack, because within just a few months he would sadly pass away. The irony was that while we were there in Rhode Island, Dr Mack mentioned that he had begun work on a new project and would be writing a book accordingly. The subject matter would be – ironically – life after death.

On the evening of the last day at Anne's house we all congregated for a final meal. I can remember clearly sitting on the pier at the end of the garden, on my own and not wanting to go into the house; I felt uneasy about something. I couldn't quite put my finger on it but there was an icy feeling of apprehension sitting in the pit of my stomach. Annie came out onto the balcony and called for me to come in because dinner was being served. Finally and rather reluctantly I paced slowly up the garden and joined everybody at the table. I continued to feel anxious without really understanding where this emotion was stemming from. As the meal progressed I began to retreat more and more from the conversation that rose and fell in jolly waves around the table. Opposite from where I was sitting was a large ornate mirror hanging on the wall. In it I could see across the hallway behind me and into the music room beyond, which also had another elaborate mirror in line with this one. I kept being distracted by something in the other mirror. At first I thought it was just a reflection, then I realised I was seeing something that was not in front of the mirror, but rather something that was in the mirror. Faces kept appearing; familiar teardrop shaped faces, with large hypnotic black eyes that stared back at me from both of the mirrors. They moved from side to side, seemingly making room for more faces that kept appearing. I sat there mesmerised by this totally abstract apparition that played out before me. As if breaking through some sort of elastic glass barrier a long slender arm snapped from the confines of the mirrored surface to be followed by another, then a face and an upper torso. With tremendous effort I forcibly broke away from this mind-boggling vision and looked down at the half-eaten meal before me. I desperately wanted to alert everybody sitting by me but felt utterly frozen to the spot, unable to move a muscle or speak a single word. All I could hear around me was a seemingly slowed-down drawl of deep resonating speech. I was terrified of what was going to happen. It just didn't feel right. Although I believed the faces to be familiar, and in that apparent familiarity there was normally safety, the feeling that I was actually experiencing in my heart was anything but safe. This was somehow different and I didn't know why. I tried to move but I just felt wooden, as if all my limbs were no longer bodily parts of flesh and blood. Something ominous was approaching, bearing down on me like a dense black storm cloud.

Raising my head a fraction, my peripheral vision sent back a picture that didn't make sense: my seated companions were still there but they were not moving, as if captured in a single frame of a motion picture, something that I had experienced many years before in the discothèque in south London on a night out with old school friends. I managed to turn my head to the left a few degrees and make eye contact with Dr. John Mack who was sitting to the left of Anne Cuvelier. In his right hand was a fork with a morsel of food stuck on its prongs, his arm caught in the motion of raising the utensil to his lips in preparation to consume his dinner. Although he was looking in my direction I felt that he was merely caught in the moment as his head turned towards me. And then the tableau changed. The fork had finally reached his lips and the tip of it was beginning to disappear into his mouth. He was animated. He wasn't motionless. Additionally, I could now see that although Anne's eyes had been open, they were now closed, caught in that fraction of a second as the eye lids blinked. Directly on my left was my wife, Annie, and I watched patiently as her lips parted and clearly altered in shape as they apparently formed specific words in the process of speaking. With enormous effort I twisted my head back round to the right, scanning the rest of the seated gathering and watching for the tiniest sign of movement. And there it was, barely noticeable unless I stayed focused on one small area, but nonetheless identifiable. The question was though, were they moving slowly or was I moving inordinately fast?

Even now this menacing presence continued to bear down on me from behind, still unseen, yet apparently inescapable. It had no tangible control of my body yet I could still feel it holding me in place, squashing me down into my seat, until finally I felt its tangible, heavy touch was upon me. It was as if a tendril of pure malevolence reached out and stabbed me in the back of the neck. I had no idea how much of this reaction was real or just my intensified expectation; magnified a thousandfold by a belief system built on the fear of the unknown and the hand of the bogeyman grabbing you in the depths of your worst nightmare.

Without thinking my head turned to see what was touching me. In my confusion, a wave of mixed emotions flooded my fragile psyche as I saw a familiar sight from my memory. A relatively small, grey, wrinkled, alien-looking figure standing at the height of my seated frame loomed by my side with its crooked arm and extended digit reaching out and touching my neck. I looked into its intensely black penetrating eyes hoping to find some semblance of recognizable kindness, but the essence of this emotion was not to be found. Although the packaging was the same the life force and its intent was most definitely not.

There was movement from around the table; several other

of these figures were moving behind and in-between the people at the table. They too were being prodded and examined closely; the difference being that they, unlike me, appeared to have no awareness of the presence of the uninvited visitors, their faces showing no sign of disturbance as their demeanour continued to reflect a gathering of people in light-hearted social exchange. Peculiarly though I noticed what appeared to be a thin, liquid, translucent, film that sat just on the surface of all I could see like a barely visible watery ripple on the face of a pond. I felt that I could reach out and touch it. As well as this watery film there was a background soundtrack that continued throughout the event: a barely noticeable, low, resonating note that fluctuated in timbre and volume. Although I could not identify it properly there was something memorable about it.

Suddenly a flash of light filled the room in a blindingly white and silver burst. It was as if no point escaped its brilliance. It came suddenly and then dissipated just as quickly.

I then sensed a new presence in the house, one of immediate and comforting benevolence. I felt the tension in my whole body flood out. I experienced a calmness that was not my own. It was as if my circumstances were being manipulated, but I embraced it without reservation. And although they physically remained, the threatening energy of our little grey intruders instantly evaporated. It was as if their malevolent presence had simply drained from the room, leaving them standing diffused and redundant. But every other aspect of this surreal stage setting remained the same. Although I felt slightly more capable of movement, the concept of actually doing so appeared beyond my ability. So I remained stationary. My attention continued to stay focused on our uninvited guests and as I stared so they became less and less present in the moment; they just started to fade away, until finally they were no longer there.

An invisible button was pushed somewhere and everybody in the room started moving and talking normally once more. One thing remained peculiar though: I just didn't feel as if I was in attendance, as if I was even there in the room. I remained seated as people around me rose and began to congregate in the hallway. I could hear snatches of conversation and it was apparent that, for some, the evening was coming to a close, whilst others were intending to go out to the garden for a midnight swim. I walked out to say goodnight and mingled with small clutches of people, singling out those that were going to bed.

I approached two people, Bob and Teri Brown (organisers of UFO Congress Conference), who were leaving early the next morning to catch a flight home. Stepping forward with the intention of saying farewell, as I knew that I would not rise early enough in the morning, they simply stared right through me. The feeling I experienced simply cut me stone cold dead. I had never felt anything like it in my life. To them, it appeared, I just did not exist. I was invisible. I wasn't being ignored, I knew what that felt like. This was different; I was a non-entity. I spoke, but they turned and walked towards the stairs.

People were helping Anne to gather chairs and generally tidy up. I observed all as the normality of events played out before me. I attempted to catch somebody's attention. I stood directly in front of Mary, the wife of one of the Weiner brothers, and spoke, not quietly but pronounced and stridently, but she did not show any response at all.

What on Earth was going on?

And then somebody was at my side. It was Little Star. Looking directly at me, I watched as her lips moved, but I could not understand what she was saying; as far as I was concerned it was an unrecognizable language. At least she could see me though, I thought! I smiled half-heartedly and gently shook my head. She smiled back, as if I had somehow responded properly to what she had been saying. She then joined Sean and they went out onto the porch.

I was standing at one end of the hallway, just by the door that Little Star and Sean had left by. To the right of me, by the entrance to the music room, sat Jack Weiner. His eyes seemed fixed on a target a thousand miles away and appeared to have him locked in a moment of frozen terror. His distant, staring eyes spoke of something that only he was apparently privy to. Further beyond where Jack sat, stood Dr. Michael O'Connell (abduction and UFO researcher). He too stared at something seemingly beyond the vision of others, whilst also breaking away momentarily and attempting to catch the attention of others standing close by. As far as they were concerned it looked as if he didn't exist. He seemed confused as if he didn't know what to do, like a child in the company of adults.

I sat down on one of the chairs that had been placed against the wall. I continued to watch the proceedings all around me. Although I could hear people as they exchanged pleasantries there was a muffled quality to the sound. Also, as one or two moved from one location in the hallway to another, their step seemed just slightly retarded as if they were walking in water.

My attention was drawn to the far end of the hallway. I narrowed my eyes in an attempt to focus on an area between the stairs and the front door. With the stained glass windows as a backdrop there was a very distinct space that just seemed to stand proud, as if a large patch of air was somehow out of focus. It must have been about ten feet in height and at least six feet wide and just hung there distorting everything that could be seen through it.

As I stared so intently, my awareness of those around me and the hallway itself seemed to diminish until it was as if I were alone and looking with tunnel vision. Within the blurriness I started to make out two figures. Although they had no discernable features they were very clearly human in outline. I could definitely make out the shape of heads with willowy torsos and spindly legs. Their height was tremendous, standing at least nine feet tall, but disproportionately narrow in comparison. I wasn't sure if it was due to the overall ambiguity of what I was seeing, but the light being emitted from these figures had an abnormal phosphorescence to it.

As if in some sort of trance state, I took a few tentative steps forwards without really thinking about what I was doing. As I got closer the figures came into focus a little more. I could see movement within their luminosity. A million minute branches of animated light twisted and turned like tiny, effervescent, wriggly worms fighting for dominance over each other. I think I moved forward once more, although I'm not certain for they could have easily have moved closer themselves. I was becoming lost in their hypnotic radiance. And then the light went out.

Due to the peculiar and unique nature of this particular contact experience, it is my sound belief that the connection continued on a metaphysical level and not necessarily with my physical body. Based

on my memory, I was then taken from the house and to a location that was 'other-worldly' in appearance, where my interaction with The Light Beings continued. My recall of what happened next remains in surreal, disconnected, bubbles of memory and the following narration is taken directly from my recorded recollection of events, spoken in the present tense, whilst in an induced state of relaxation.

"With regard to where I now found myself," I said: "...it looks like ice around me. Stretching out before me... an ice tube, crystal ice. I'm in a long tunnel. I'm just sitting here, but it doesn't feel like I am waiting for anything, it just feels like this is where I am supposed to be. I don't know if I have gone somewhere or I am waiting to go somewhere. It's like a long tunnel; it's made of ice, it's really smooth."

I went on to comment about how I felt. "It just doesn't feel like I'm me anymore. It doesn't feel like I'm in here." This feeling of being disconnected from self continued throughout the experience. "I've come from somewhere and now I'm here and I'm going to go somewhere else, but I can't even see my body. I can't even feel it."

My description continued in disjointed bursts of senses and observations. "I don't even have to move my head to look around; I can already see all around without moving. It feels like there is a pressure change. It feels like there's something coming. I'm not scared though, nothing can hurt me the way I am. I'm in between, in between two things and I'm just waiting. My life is behind me; my life is where I have come from. Something is tinkling... there's a tinkling sound, like very delicate little chimes and it feels like everything is changing around me and it feels like I am no longer where I was. It's not like there is a behind me, or beside me, or even an up and down. I can see all around me. I'm just aware of everything all at once. It feels like something is pulling me now."

I then experienced movement as if I were being *transported*. "The pressure, it...it feels like something is pulling me. It feels like I'm being pushed and drawn at the same time." I could clearly see my surroundings in extremely close detail. "You can see into the walls, they're made of ice or crystal. Whatever it is, you can see into it.

There are 'things' in it that are moving. It's like little trees without the leaves, just fragile, spindly branches and lots of branches coming off those. It's all around me and the walls move off and are bending and you can't see the end of them."

I attempt to describe how the motion is making me feel. "It's not as if I am travelling somewhere. It's not like I'm in a car, travelling from A to B along a road. It feels like where I am is moving; the actual space that I'm occupying. I'm taking everything along with me; as if everything around me is a part of me, a part of who I am. It's like I extend into everything around me and we are moving towards something together, but it's not like travelling."

It's at that moment that I am aware of a presence close by. "There's somebody here with me. I can't see them though. I can see all around me, but I can't see them. They're not bad though...their presence makes me feel good. They're making me feel very safe. I feel very loved."

It felt like I was on the move again. "I'm somewhere else again. It's like the tunnel is changing...it's like there's a sky, there's trees and a hill in the distance. It doesn't look normal though. The tunnel has altered, it's morphing into different shapes. There's a path going up a hill and there are some trees and a mound. I can see a great expanse of sky, but I can't see the sun."

"Once more I have an overwhelming sense of a presence close by, but this time I can actually see them, "I can see something at the top of the pathway. There are two figures standing there. They're beginning to move towards me now. They are so huge, I can't believe that they are actually people." I feel a sense of awe at the sight of these beings and not just because of their size. "Light is just bursting from them. They don't look like normal people, but they are people nonetheless. And the light is coming from them, it's blinding. As they move closer I can see right inside; they're made of the same thing as the tunnels. Tiny, crystal branches that are constantly growing and fragmenting, twisting and turning. There is a distinct shape of a head and shoulders, and there are arms, but they're not moving and remain motionless by their sides. I can see legs, but they're not clearly defined, just a shape that suggests legs. There's a slightly darker outline all around the body. It feels like the outline is holding everything in."

I continue to be stunned by the size of these figures. "They're huge, I can't believe it, and they must be two or three times taller than I. They are so alive, I can just feel it, I can feel their life force. The outline of their bodies is just like Vincent van Gogh's paintings; he puts a dark line around everything he paints, which seems to hold in all the bright, vibrant colours. And all the crystal shapes just keep flowing around inside."

Somehow or other we appeared to be getting closer and closer to each other. "The nearer they get to me I feel like there is a feeling of liquid rising up inside of me, filling out from my toes and going into every part of my body. I feel warm all over. They're right in front of me now. I feel overwhelmed by their presence. The flowing liquid crystal of their body...it's all I can see. I stare at the face, but I can't see anything initially. As the light inside rolls and flexes, faces keep coming to the surface and then disappearing. I don't recognise any of the faces, but they are all human and look very loving. They all look at me as if they know me; it's very comforting. I can hear a voice, it sounds like they're saying that 'this can be whatever you want it to be."

I try desperately to describe these Beings in a way that adequately describes how they appear to me. "I don't know what you would call these people? They are just Beings made of light. No, no...it's more than light...it's like something organic; it's just so alive. It's like imagining a large crystal and watching it become liquid". I felt so frustrated by the limitations of language to describe these remarkable Beings. My attention to the make-up of these Beings was momentarily distracted as I looked beyond them. "Everything behind them is starting to change. It's rolling and shifting; it's not what it was...I can see that there is movement behind them now. I don't know what they are, but there are solid shapes forming."

The thought that I am somehow connected to the Beings on a deeper level continues to press upon my consciousness. "I get the impression that these Beings have actually turned around. It feels like they have turned around but it doesn't look any different. I can't see the faces appearing anymore. It just feels that we are somehow linked together. Just like before, when I felt like I was moving I'm sort of travelling again. I am 'becoming' somewhere else. Does that make sense? I am becoming somewhere else and I'm part of these two Beings now and we are moving into these other shapes. There's movement, as if these shapes are alive. I believe now that these shapes are actually buildings of some sort."

My description of my surroundings continues: "I don't know where I've come from. I don't understand that. I just seem to know that I'm here. There's a building and the side of it appears to be in a step formation from the bottom to the top. It is very, very tall. There's a doorway in the corner of it and I'm supposed to go into it. I'm not fearful, but I am apprehensive, but these two Beings are somehow wrapping something around me." I immediately feel more relaxed with what is going on. "It feels like they have extended themselves around me. The apprehension I felt isn't there any more. I'm going into the building now. It's very cool in here. The walls are not like the surface outside; there's no movement in them, they are solid like marble. And they're smooth and cold. How I know that they're smooth and cold, I don't know, because I don't seem to be able to touch them." I ponder this dilemma for a moment and then continue. "On the right-hand side there's a wall and it looks like there's an opening with a light inside it. There are long windows in the marble wall. Well, they're not windows really, I guess they appear to be openings and in front of them is something like a big stone table." Once again, I appeared to falter over the correct use of words to describe something. "I don't want to say the word... altar...it's a table, but it is for 'presentation'. You don't put something on it; you 'present' something on it."

The organic nature of what I'm seeing continues to astound me and yet another shift occurs. "The walls are convex; they're expanding and becoming something else again. I seem to be just observing this and accepting it, and it just seems like it's all right and it's normal. The walls are becoming the tunnel again and the two Light Beings are either side of me. I just feel like I am 'wrapped' in them. We're moving again. I think I understand it better now; it is as if wherever we are, we take that with us to where we go. It's not as if we're just independent of our environment and we travel along a road; we seem to take everything around us and we stretch it to where we go. I don't know, but that seems to be the best way to describe it: where we are becomes where we go to."

"There are lots of people here now (they appeared to be Human Beings dressed in a selection of brightly covered one-piece coveralls). They're seated in tiers, like a half circle of banks of seats rising up. I don't know if they can't see me or if they are just not bothered, but I'm standing in front of them and they're paying no attention to me. The movement inside The Light Beings, the movement in their bodies, is starting to slow down; it's like they are starting to become solid in a way. The inside is becoming a sort of darker hue and the movement is slowing down as if they are taking form in a way."

I begin to talk about what had happened in the house previously, when I saw The Greys coming out of the mirror. "When I saw The Greys before, I immediately believed that they were the ones I've had contact with before. Although they looked pretty much the same, with some differences mind you, their energy was most certainly very different. Their purpose was to cause an imbalance between everybody in the house and their prime objective was to affect Little Star and me. She knew they were there too. I was so scared. I just called out, in my mind, for help. And that's when The Light Beings first arrived; they didn't allow anybody to see them at that point, but they were there. They're not allowed to make The Greys go away by force, they can't tell them what to do as such, but they can persuade them to leave. You see The Light Beings are *very polite* (as we would understand it) in how they deal with situations like this. They are so strong that you can't refuse them, but having said that, they can also be quite vulnerable. In fact, they carry some sort of device when they are resonating in a physical form, as explained to me by Jason Andrews, son of Anne Andrews, whose contact experiences are covered in her book *Abducted*. The Light Beings have to be really careful when they come into direct contact with us, human beings that is, because we can *infect them*, so this device protects them from our energy. Apparently we can be like a virus; if we infect one of them it will infect them all. So The Light Beings approached The Greys and asked them to leave."

I returned to the description of The Light Beings. "I can see that thing around their waist quite clearly now. It's to stop us, or anybody come to that, from infecting them. It looks like it's attached to some sort of belt around their waist. I would say it's about eight inches long and had rounded edges. It has a sort of brushed metal effect to it, quite shiny. The Light Beings look really solid now, they're still shining in a way but that sort of internal sparkling effect has gone. I can still see pulses of light going on though, like ripples of electricity in a way, I can see it throughout their bodies."

I began to feel uneasy. I had a sense that I had to leave and the process had already begun. Although I could still see my surroundings it was as if I were being forcibly disconnected. After my strong sense of being wrapped up in the energy of The Light Beings, it now felt that I was being let go, I was being released. I felt extremely reluctant to leave where I was. I was overwhelmed with a great sense of melancholy. The simple fact was that I believed I belonged there. However much force I could possibly muster, I knew it was pointless to try to resist what was happening. Eventually, I just relaxed and, in doing so, I was immediately caught in a strong current that seemed to thrust me forward.

The transition was gradual, but soon the sense that somehow I was once again becoming heavier and denser returned. I could feel my physical body once more. Before me was a familiar scene: it was night-time and there was the brightly-lit swimming pool in the garden of Anne's house in Newport. In the water, laughing and splashing, was a group of people; amongst them I could see Annie, my wife. It seemed as if I was above them all and I could see a bright, unnatural, green light being reflected off the surface of the pool. I believe that The Light Beings had done something to the water in the pool. They told me, "It had to be different". I was still above them, but it then felt as if I was observing them through a window. It was as if I were gliding towards the house, coming in from the sea, very, very slowly. In an instant, I was walking on the grass by the side of the pool; I could feel the wet blades of grass beneath my feet. I watched with envy everybody playing in the pool, but I knew that I would be unable to join them; as I was aware that they would not be able to see me.

"When you open your mind to the impossible
sometimes you find the truth."

Walter Bishop

Experience: Our First Trip To Anne's Home

Newport, Rhode Island – July/August 2004

Part 3: The Morning After

Upon waking the next morning I felt greatly hung-over, although at the time I was still not drinking alcohol at all. Bleary-eyed and moving with leaden feet I struggled heavily into the breakfast room and plonked myself down. Normally, even at this early hour, the assembled guests would be involved in highly charged discussions but this morning was very different. Although dialogue was being exchanged there appeared to be a somber element to the proceedings. Having said that, there were a couple of individuals who were their normal effervescent selves but they seemed unable to transfer that feeling amongst the rest of us.

Eva was the first one to talk about what happened the night before. "Something strange happened last night," she said gravely. All heads turned towards her. "Did anybody feel anything odd at the end of the evening?" she continued. Nobody answered. "I just had the strangest of feelings when we were finishing the meal last night that there was something ominous in the house, okay." She certainly had everybody's attention now. 'And when I went out into the hallway to say goodnight to the people who were going to bed...I don't know, *it just felt as if they were ignoring me."*

The memory of last night started to return in thin, wispy, currents of thought. Eva's words were certainly prompting my own recollection of what had happened.

"I stayed around for a little bit, okay," Eva continued. "But I just felt so uncomfortable – as if nobody really wanted me around – that I eventually just went to bed."

I didn't say anything; I just did not have the energy somehow to form words. I felt completely washed out.

Mary Weiner slipped silently and unobtrusively into the room, poured herself a coffee and sat down. A few minutes later she was joined by her husband Jack. They sat quietly together and appeared, to me anyway, to be very *tiny* somehow. Clearly reluctant and in a tiny voice, Mary Weiner, began to speak. "Did anybody see the strange lights outside last night?"

Heads swivelled away from Eva who had carried on talking, and were now focused intently on Mary. Looking down at the breakfast plate before her Mary continued, but it was obvious that she was finding it difficult to share this with us.

"When Jack and I went to bed last night, it wasn't long before he went off to sleep, but I turned restlessly in bed for ages. Then I could feel as if something was in the room." She shook her head. "The room was really dark but there was something moving about *inside* the room. There was loads of... *activity.*"

Jack sat beside her; he had a hand placed on hers. She started to speak again, but her voice began to breakup and so Jack continued on with the story.

"I'd been asleep a while, but then something woke me. I got up and went to the bathroom. When I got back Mary wasn't in the bed; I don't know if she had been there when I got up or not. I turned on the lamp and I could see Mary's head bobbing about at the end of the bed – she was on her hands and knees by the door looking out of the window. She kept on insisting something was trying to get in." Jack looked round at Mary and smiled sympathetically. "I was staggered! Who would be trying to get in here? I asked her. What on Earth are you doing on the floor?"

"You know where our bedroom is," Jack asked, looking about the table. A couple of heads nodded. "It's downstairs and backs on to the garden," Jack explained. "There's a sort of porch just outside one of our doors that has a door to the garden and a door to Steve and Annie's room." More heads nodded. "Mary was on her hands and knees pulling back the curtain looking out to the garden, so I helped her get back to her feet and put her back to bed, but she just kept saying that *something* was out there."

Mary came back in at this point. "When I was lying in bed and I really thought somebody was in the room, but eventually it just went still. Then I could see a light in the porch and then that went out, but it came on again outside our bedroom window. It was a sort of interrupted light, sort of flashing but not really, and for some reason or other I kept telling myself it was the porch light, although I knew it wasn't. Then I convinced myself it was the moonlight shining through the mast of a boat, which is completely crazy I know."

Mary stopped and looked down again. "I was absolutely terrified. I made Jack get up, but he wouldn't go outside. We just lay there in bed and held each other."

Then Jack added: "I was lying on my left side and watching a somewhat narrow – about five inches wide – very strange pattern of multiple, horizontal bands of coloured light modulating up and down the entire height of the back wall of our room and thinking that *something didn't seem quite right about it*, but then again, *it must be the light from the lighthouse out on the island opposite Anne's house.* Eventually, I fell back to sleep and stayed that way until this morning. Mary and I both washed up and as we dressed we discussed the strange night before and what had been going on in the porch. Mary went up first, then I followed on. On my way up to breakfast I took note of the very large, full and leafy bush located right in front of the entrance to the porch that would certainly have blocked any light from entering our room. Hmmmm, I thought, if it wasn't the lighthouse, then what *was* making that strange light on the wall?"

It was clear that whatever had happened the night before had had a profound effect on Mary and she rose from the table and left. Jack soon followed, apologizing to everybody for leaving so abruptly. Additionally, Mary and Jack have experienced ET contact at their own isolated home in the mountains. In a recent e-mail from Jack (May 2010) he states the following:

Hi Steven. When I met you and Annie at 'Anne's' several summer's ago, I knew immediately that there was something unique about you both. Being easy to converse with and very likable company, we had some very interesting and enjoyable conversations concerning the apparent similarities and differences of our UFO encounters, how our lives were being changed by them, the ramifications of 'going public' and what life was like in the U.K.

Jack goes on to recount and comment about the events of that strange night in Newport, which I have documented in the above narrative, and then he closes with the following:

There was absolutely no doubt about it... something in the order of 'Extremely High Strangeness' happened to us all right in Anne's house that night. Later that day, you and Annie left in a mad, frightened rush and that was the last time we saw you. All who remained at Anne's felt overwhelmingly concerned for you, uncomfortably vulnerable and more than a little scared.

Later, we were all relieved to hear that you would be 'OK'. Although there were a few of us who knew all too well what that meant though: 'Would be OK' can sometimes take a long time to materialize. Most people are simply not aware of some of the harsh realities that one must grapple with during, and especially after, an abduction event. And, as many well-conducted studies have revealed, the contacts can continue indefinitely. As you know, abduction by some kind of 'Alien Presence' is not something that can ever really be forgotten, and those memories of the experiences, can haunt and tantalize us until our dying moments. It's certainly not a walk in the park. No matter who you are.

I'm glad you've written your book Steven, because now the whole world can read about your totally incredible and true story of contact with an Extraterrestrial Intelligence and how it influences your life. As weird and unfathomable as some of our abduction experiences are though, at least 'They' haven't tattooed our bodies with 'Crop Circle Designs!' NOT YET ANYWAY!

Best Regards,
Jack Weiner

Returning to what transpired at the breakfast table after Jack and Mary left, I recall the following conversation:

"What about you, Steve?" Eva asked. "You don't seem your bubbly self this morning and I know it's not a hangover."

Not yet comfortable with my own experience – or even that clear, come to that, with what exactly had transpired – I said, "I'm just tired I guess, but there was definitely a strange atmosphere going on last night. I don't know what it was."

"And another thing Steve, where did you get to? I remember seeing you at dinner and then at the end you weren't there," Eva quizzed.

Just then two latecomers arrived to the breakfast gathering. Little Star and Sean filled their coffee cups and sat down at the seats left vacant by Jack and Mary. The room was silent. Our new arrivals studied their coffees intently as they continued to stir their dissolving sugar way beyond the need to do so.

The best description of Little Star in a social situation is *sparkling*; her laughter bubbles like a gentle stream and her positive state of mind is really quite contagious. Therefore, as she sat there obviously determined not to engage, it troubled me greatly. I felt tremendously protective of this tiny fragile figure and clinked my spoon against the side of my cup, in doing so, catching her eye. I flicked my eyes to one side, stood and walked out into the hallway. A moment later, Little Star and Sean appeared.

Sitting down in the music room I could see Little Star's eyes welling with tears. Sean had an arm placed sensitively around her waist as they cuddled up on the couch. Annie then came in and sat down beside me.

"What's going on sweetie?" I asked Little Star, as she produced a tissue from her pocket and dabbed away at her tears.

The corners of her mouth turned down and she shook her head gently. She looked so vulnerable, I just needed to wrap her up and protect her.

"I've got so much I need to speak to you about, but..." she said falteringly. "Do you remember me speaking to you in the hallway last night?"

"Yes."

"I sensed something in the house you know. There was something really ominous. Sean and I were going to join the others and you were just standing there in the hallway looking completely dazed. Do you remember me speaking to you?"

"You *definitely* remember seeing me and speaking with me, do you?" I asked, apparently too sharply.

Little Star looked at me strangely. "Of course I do." She waited a second. "Your eyes were like saucers. The pupils were totally dilated and very, very dark. I asked you to come and be with us, but you didn't even reply, just kind of vaguely smiled at me. You looked really peculiar; I was a bit worried."

"Yes, I did feel really weird as if everybody was ignoring me."

"I don't understand. What do you mean 'everybody was ignoring you?'"

"I tried to say goodnight to a couple of people, but they acted as if I wasn't even there."

We just stared back at each other, not really knowing where to go next. Little Star looked round at Sean and he nodded as if confirming something.

"I sensed The Greys were here last night," Little Star said finally. "But the thing was, they felt different, they felt bad." She blinked a couple of times and as she did so tears rolled down her cheeks. "I didn't like them; they seemed really uncaring. But...there were..." she hesitated, collecting her thoughts. "There were 'others' here too," she said, looking round to Sean as if for further confirmation. "We call them the '*Light Weavers.*"

My eyes must have widened, because Little Star just reflected back the gesture.

"I felt that they tried to intervene. They can't stop The Greys, but it felt like they were trying to get them to leave. *The Light Weavers are very polite you see.*"

My eyes must have looked as if they were about pop out of my head because Little Star giggled, clearly finding amusement in how my face must have looked and also releasing some pent up tension from recalling her experience.

Light Weavers. Intervening. Very polite. Oh my god!!!!!

"When we finally went to bed," Little Star went on, "I was so apprehensive that the bad Greys might still be around that Sean had to go with me to the toilet. Hah!" She looked a little embarrassed. "I spoke to the Light Weavers and they put a protective energy around our room, it was like a cobweb of blue and white light."

I told them about parts of the experience that I could remember. Annie was hearing this for the first time and clasped my hand tightly into hers. I explained that I thought that The Light Beings/Light Weavers hadn't actually taken me physically and Little Star agreed that they don't operate on our three-dimensional level, so they would have taken my *light body* to a different location. We all agreed that the location could possibly have been their 'home planet' or the 'vibrational location' where they exist.

Addendum

One of the guests at Anne's home that week was Dr. Michael O'Connell who is, as I have previously stated, an alien abduction and UFO investigator as well as being a graduate of Harvard, a clinical psychologist and a veteran of many abductee regressional hypnosis sessions. After he heard that we had had experiences on the last night, he shared with us his own memories of what occurred for him.

'I had an odd experience with Steve at Anne's house. Our last dinner together was Thursday evening and some people had to leave at 4 a.m. to catch planes, so there were some goodbyes being said after dinner. About 15 people were standing in small groups all over the dining room and hallway and music room in conversation and saying goodbye. I wandered around the groups but was totally ignored. I didn't feel welcome in any of the groups. No one looked at me or made any gesture to bring me into the group. That struck me as most unusual as we are like a big family and that behaviour had never happened in my 10 years of going there.

I went out to the large porch at the back, overlooking the ocean, for someone to talk to, but it was empty. I came back inside and saw our hostess and another friend putting the folding chairs away. The dining room is vast so its like being in a huge room with the hall combined. I thought I

*should help them, since no one else was, but for some reason
I didn't do that. That seems odd to me, I wanted to help, but
I seemed frozen, just watching.*

*A week later in conversation with Sapphire, I related my
experience of going out onto the porch and looking for people
to talk to and finding it empty. This had prompted me to just
go to bed. The time was 9.45 and I had never gone to bed
that early before! Sapphire shocked me by saying that on the
contrary there were several people on the porch and none had
seen me come out and look.*

*The following morning, Friday, Steve seemed to be
daydreaming and distracted at breakfast, and he said the
last thing he remembered from the evening before was sitting
on a folding chair in the hallway after dinner, but then
waking up in bed the next morning. However, had his body
been sitting there, I would have certainly stumbled over him
on my way outside to find someone to talk to. And I would
have certainly stopped to talk to him, instead of looking for
someone else to talk to. We subsequently discovered that he
had a major contact experience during the dinner, the night
before.*

*Jack and Mary who were in the bedroom next to Steve
and Annie, had also been affected by the experience. Jack,
by being 'switched off' in some way, again in the hallway,
and Mary, by being terrified by the bright white light outside
Annie's and Steve's room. Jack had found her hiding behind
the settee in their room and peeping through the curtain out
to the bay.'*

Michael showed me a great deal of love and care after the event
at Anne's and paid a visit to where I was staying to offer counsel and
support. Subsequently, he took me through an hypnotic regression
session to add depth and clarity to my memory of the contact experience,
which has brought a great deal of understanding and healing to what
occurred for all of us on that last evening at Anne's home in Newport,
Rhode Island. I owe Michael a great debt of gratitude for all the help
and support he has given me over the years.

"When your inner eyes open,
you can find immense beauty
hidden within the inconsequential details of daily life.
When your inner ears open, you can hear the subtle,
lovely music of the universe
everywhere you go."

Timothy Ray Miller

Experience: Nobody Else Can See Me!

New York City – August 2004

In the summer of 2004, Annie and I were on holiday in America and we were staying in a really funky hotel called *The Hudson*, just off Columbus Circle and across the road from the southwest corner of Central Park in Manhattan. We had just come down from Rhode Island, where we had spent the week at Anne Cuvelier's home in Newport. In truth, my head was spinning because I had just met Sapphire/Little Star for the very first time on 'terra firma'. We were waiting for Anne C., Sean and Sapphire to arrive, as they were coming to join us in the Big Apple for an overnight stay. Annie and I had been down to the Strand Bookstore that morning, to purchase a very important and relevant book, which I would inscribe and give to Sapphire as a gift upon her arrival: the famous and enchanting story by J.M. Barrie, *Peter Pan*.

Just around the corner from the hotel, right on Columbus Circle was our corner store. In reality it was quite a bit more than that: the Whole Foods Store is one of the most impressive supermarkets we have ever shopped in and contains a selection of wholesome organic food way beyond anything I had ever imagined. One morning we found ourselves purveying their organic fare in preparation for a picnic in the park and once chosen I stood in line waiting to pay, whilst Annie was still browsing.

Due to the huge size of the store and the amount of customer traffic they have, there is a very professional set-up for queuing and purchasing. As I stood there, shuffling along, watching the faces of the other customers, I saw a woman strolling casually across the store. She was very trim, pretty and well dressed with refined features, dark hair and a distinct calmness to her nature. As if instantly aware of my gaze she looked directly at me, although the distance between us must have been at least fifty feet and was interrupted by other people walking and standing. It was as if our eyes zeroed in on each other and the distance seemed to disappear so that it felt as if she was right there before me. I felt startled and relaxed both in the same measure. I was unaware of my environment – there were now no other shoppers or any noise coming from around me – all I could see were the eyes of this woman; they held me to the spot, to the moment and I had no sense of time.

Then she spoke. Her lips did not move, but still I heard her voice. Softly and with precise enunciation, she said, "Nobody else can see me, please do not bring attention to me."

The words were not aggressive or forceful; it was more of a plea than an instruction. As if in response to some unheard level of communication from myself, her mouth moved with just a hint of a smile and her head tilted with a suggestion of confirmation. And then she moved on. Gliding with effortless ease, she was finally obscured by the dual escalators that rose and fell in the middle of the room. I did not see her reappear.

On Annie's return, I immediately told her what had happened. She smiled, her eyes sparkling with apparent glee as she told me of a strange incident one of her friends had once had whilst travelling on the London Underground train system in central London. The woman in question had boarded an underground train and sat herself down in an empty seat. As she was getting comfortable, straightening her clothing and retrieving a book from her bag, she noticed a man sitting opposite who took her eye. He was a man quite nondescript in many ways although he had the most piercing eyes of a bluish/mauve hue.

She stared back silently. Then she heard a voice in her head. It said, very precisely and clearly, "Please don't give me away. You can tell that I am different. I am not from this planet. Please don't give me away."

The man smiled as if requesting some sort of confirmation. Annie's friend smiled back and the man showed a clear sign of relief. At the next stop, the man rose from his seat and alighted. The woman did not see where he went and has never seen him again.

The similarity with my experience was clear. Had I just encountered an ET of some description? Was this an example of what has come to be referred to as a 'Transgenic Being' – a genetic mixture of human and ET DNA? Or had we fallen into the trap of believing that ETs should only have an alien appearance? Maybe this was exactly how most ETs actually look – *human*!

Food For Thought: Intelligence Extra – Who is Watching Us?

Written on Wednesday 30th December 2009

O ver the last few weeks Annie and I have been following a fascinating natural history series on BBC 1 television, narrated by David Attenborough and tilted *Life*. Each episode has covered a specific element of life on this beautiful blue planet of ours. This evening's programme was called 'Plant'.

The overall production of the series has really been quite outstanding. The camera work and especially the sound surpassed anything that I had seen before. I am normally touched by programmes of this type, but this time I was moved in a way that found me welling up with emotion that surprised even me. And last night was no exception.

Considering that the subject was plant life, I naively believed that it would not reach the emotional levels achieved by earlier programmes on animals. The production team went on to show that plants are just as much alive and animated as the animal kingdom. Using advanced filming techniques to record the slow growth patterns and then replaying them at an increased speed, brought their subject matter to life in a way I wouldn't have thought possible. Watching from an unavoidable anthropomorphic viewpoint, I saw arms and legs twisting and curling, bodies and heads pulsing and turning, and human intentions played out before me by a form of life far different from our own.

For the first time in my life I was actually seeing plants as intelligent, and maybe even sentient. Was this just my point of view, based on how it was being presented to me? Or had a veil been lifted to enable me to see a hidden truth?

But then I had a concern: up until this point I had gone along with the theory that plants *might* be intelligent, based on the odd bit of information I had gleaned here and there, plus my intuitive sense of it all. What had pushed me beyond these original, and I might say restricting, thoughts was watching the plant's growth and activity speeded up to a rate that fell into line with my own, thus giving them the appearance of intelligent behaviour. But before, due to their slow movement, I hadn't thought that way.

Now, I would like to propose the idea of an advanced life form operating on a higher frequency rate than ours, but still being able to observe us in our living state. Would they consider us not intelligent, not sentient beings, just as I previously had considered plant life not to be? We look out into the cosmos, searching for signs of life based on our concepts of what they might be, but have we started our journey from the wrong place? We believe that the human race is the most intelligent form of life on this planet, but would an alien life form consider that to be true? Based on their interpretation of intelligence, maybe they have already looked and, not finding the credentials that fit their concept of sentient life, moved on? Or consider this: might they have already made contact with another species of life on this planet, one that successfully ticked all the boxes?

*"If the whole truth does not reside in any one location
you can be sure that the pieces to the puzzle are
waiting elsewhere for you to put them together."*

Steven Jones 2010

Social Event: My First Presentation at Laughlin

UFO Congress Conference, Laughlin, Nevada – February 2005

When Annie and I were at Anne Cuvelier's house in Newport, Rhode Island for the very first time in August of 2004 – the year I met with Little Star outside of an ET environment – two of the guests were Bob and Teri Brown, the directors and organisers of the world famous UFO Congress Conference, based in America. It was around this time that I made a conscious decision to start talking publicly about my contact experiences: I had already spoken in London at BUFORA (British UFO Research Association) and once at the Intruders Foundation in New York the year before, but now I planned to document my experiences and seriously attempt to find a way to articulate what I believed they really meant, not just to me, but to everybody.

I had put together a sort of personal Experiencer Curriculum Vitae, so when I met up with people like Bob and Teri, I could present both my experiences and myself in a professional manner, so I would be considered a candidate for consideration at further conferences.

Annie and I found Bob and Teri to be really exceptional people, based on their ceaseless and untiring work to create a platform to allow extremely important and vital information to reach as many people as possible. Having already attended their conference the year before as audience members, we had been astounded at the

Conference's level of professionalism and expertise. During the week at Anne's home, we shared many special hours together with them discussing the UFO phenomenon late into the warm summer nights out on the porch. From that moment on Bob and Teri became very close friends, and even though Teri has now left us, Annie and I still love them both very dearly.

A few months after we said our goodbyes I received a letter from Bob and Teri asking me if I would consider presenting at the Laughlin conference in February of 2005. Although they had said they would consider me for future conferences I really didn't believe I would receive a call so quickly, if ever. Looking at the list of illustrious presenters they had lined up at the conference in the past years I felt extremely honoured by their proposal. I just hoped that I would be able to meet the very high standards already set.

I spent a lot of time putting together my presentation, as I didn't want it to be just a regurgitation of the events involving my experiences. I have always felt that the ET contact meant something more to us as human beings than what was being presented by the media. I always felt as if the ETs were holding up a mirror for us to look in and that it was vital we all considered our individual roles on this planet and our interaction with each other. If the contact wasn't exclusively about this, I still felt that it was an extremely important by-product. Part of my presentation needed to speak directly to my fellow Experiencers: people who might still be in the process of working through the fear factor, and my words might hopefully play a part in their own self-empowerment. I also wanted to reach the investigators and the science community; I wanted them to consider that to move their research forward they might want to talk to us, the Experiencers, thus getting information 'straight from the horse's mouth', as the saying goes. They can speculate upon strange lights in the sky, ridiculous theories banded around by the press, dead-end hypothesis and self promoting conjecture, but the Experiencers are in direct contact with this remarkable alien consciousness, so why not start their exploration and questions with us? By making this statement, I am not putting Experiencers on a pedestal or attempting to create some sort of ridiculous hierarchy, I am just stating something plain and simple…and blatantly obvious: the ETs are talking to *us*, so to be able to find out about them, you need to ask *us*.

Attending the conference as part of the presentation team were Ann and Jason Andrews. We had become very close friends with the Andrews family after I met Ann and her husband Paul at a little, one-day, UFO conference in East London a year or so before. The whole family have an ongoing history of ET contact and paranormal experience, with the ET contact focused on her son Jason, which Ann Andrews has written about in several books that document their experiences: *Abducted: The True Story of Alien Abduction in Rural England*; *Walking Between Worlds, Belonging to None*; and *Jason, My Indigo Child*. I could not even begin to tell you about the unimaginable things that they have all gone through, so I highly recommend their books to you to find out more.

Jason and Ann had made the presentation just before me, so it was with more than a little trepidation that I took to the stage, hoping to be at least half as good as they had been. What came as a surprise to me about this sort of situation is that with the spotlights shining directly into one's eyes, it is virtually impossible to see anybody sitting in the audience at all. Therefore, as I began talking I hoped that there were still people in the hall. As is my wont, I have been known to crack a joke or two now and again, so I thought I was on pretty firm ground when I made a crack about the ABC News anchor man Peter Jennings, whose show had just recently carried out a 90-minute hatchet job on the world of UFOs and had attempted to dismantle and ridicule the enigma of alien abduction. Jennings' research crew had come down to Anne Cuvelier's home the year before whilst we were all there and had interviewed a number of us, myself included. Little Star had warned us all that nothing good would come of it and she was proved right. Anyway, I thought that Peter Jennings had a lot to answer for, so I made a joke at his expense:

"In the early hours of this morning, a small UFO landed in the car park in front of the hotel staging the conference. An alien Grey got out of the craft and came into the reception area. He went up to the desk and handed in this envelope (at this point I produced a white envelope) and asked for it to be read out at today's presentation. (I now opened it and took out a single white piece of paper!) 'There doesn't appear to be much written on the paper, but it is addressed to Peter

Jennings of the ABC Network. (I wait a second or two, clearly milking the comic moment) It just says, 'We know where you live'."

Well, I got a laugh and I was pleased to find out that there were people in the audience and I wasn't on my own – in fact there were well over 700 people sitting in the darkness.

I felt the presentation went down well, so did Bob and Teri Brown, so did Annie and our friends, so did the wonderful members of the audience who came up to me afterwards to offer their congratulations on a job well done. I still wonder if I achieved what I set out to achieve; I hope so. I have never found out what 'they' thought, the ETs; maybe I should have asked them.

"We all know that UFOs are real.
All we need to ask is where do they come from,
and what do they want."

Captain Edgar Mitchell, Apollo 14 astronaut
and Director of The Institute of Noetic Sciences

Experience: Silver Balls in the Big Apple

New York City – Summer 2005

New York City in the summer is nearly unbearable because of the heat. No wonder in the film *The Seven Year Itch* all the wise New Yorkers leave the city for the cooler coast, whilst Marilyn Monroe's character remains, but very wisely puts her underwear in the ice box to reduce her body temperature. I remember watching this movie and not believing that the heat could possibly be that oppressive, but I can assure you that it is. On our visits to Manhattan during this period, Annie and I would take to the streets (we are great walkers when we go to the Big Apple) and we'd eventually feel as if we were melting into the sidewalk. One of the great 'relievers' from the heat remains diving into a wonderfully air-conditioned cinema up on 42nd Street, which we would do most days. On one particularly sweltering occasion we spent the whole day there, ultimately taking in three films. Therefore, in August of 2005 the weather was no different really, except that they were experiencing a higher degree of humidity than usual. Oh deep joy!

Annie and I had had a wonderful experience whilst in New York this time round and were planning to catch a flight home from JFK on Sunday evening. As is our custom, we like to squeeze every last drop of delicious juice out of our time there, so in the morning we went down to Battery Park to wave goodbye to the Statue of Liberty, then watch the children dance about in the fountains that shoot jets of water out of the ground and then on to the Financial District for a stroll amongst the silent skyscrapers. As we walked, hand in hand along Broadway, after one last browse through the 18 miles of bookshelves in the Strand bookstore, I happened to look skyward – something that is not normally done as there is so little sky to see when surrounded by pillars of steel, glass and concrete – and I noticed in the slim band of blue sky that ribboned along above us in a disjointed stream was what appeared to be silver balls, reflecting the glorious morning sunshine. After observing these sparkling orbs for a minute or two, we realised that they were in fact stationary – not a normal characteristic of balloons, which was how we had initially identified them. As we counted, we both agreed that there were about fifty of these spheroids, in an apparently disconnected chain, hanging in the part of sky that was observable between the buildings.

I attempted to take several photographs of them, but to no avail; because of their height and the restrictive limit of my lens it was virtually impossible to see the silver balls via the small LCD screen on the rear of the digital camera. Even so, I snapped away, hoping that I might be able to enlarge the image once I downloaded them onto my laptop when I got home. (Update: this procedure never produced anything of any worth – when I enlarged the image there appeared to be nothing there except a clear blue sky). Unbelievable as it seems to me now, Annie and I agreed that we had better move on because we needed to get back to the apartment for some last minute packing. So we left our unidentifiable flying objects and caught the subway back up to the West Village.

When six o'clock came we took our suitcases down to 6th Avenue and hailed a cab to take us out to the airport. This is always a somber moment as we head out across the bridge and look back at the sublime skyline of Manhattan, but this time we had company! That certainly took our attention away from our final gaze at the Big Apple. There above us and just to the left side of the Yellow cab, was one of the silver balls we had seen earlier. This time it was much lower than it had been previously and appeared to be in motion. In fact, if anything, it was, we both agreed, keeping pace with the cab. As we sped along Flatbush Avenue through downtown Brooklyn, the silver ball stayed constant and did not alter its size, indicating to us that it was clearly maintaining its speed. This was such an exciting finale to our trip that our usual solemn spirits at this stage of the journey were lifted tremendously. Our unidentifiable companion stayed with us all the way to the airport, never changing its appearance at any time but simply remaining there as if it was attached to the cab by some invisible cable, only losing it as we finally slowed down for the entrance into the main airport.

Our sense of it was that the silver balls we had seen floating above the Financial District earlier that day were somehow connected with this new apparition. Clearly, this was an intuitive sense but one that had proved reliable on many, many occasions. Was it communicating with us? Was it saying goodbye? We both felt that it was definitely intelligently controlled – there was no doubt about that as far as we were concerned. We could feel the energy of this orb and the connection – although unseen – that it had with both my wife and myself.

"Extraterrestrial contact is a real phenomenon.
The Vatican is receiving much information about extraterrestrials
and their contacts with humans from its Nuncious (embassies) in
various countries, such as Mexico, Chile and Venezuela.
Of course, the aliens and their vessels do quite definitely exist."

Monsignor Corrado Balducci (Vatican theologian)

Experience: Four Around the Table

Location unknown – November 2005

Friday 18th of November was like any other day really; apart from the strange, yet familiar, pressure I was experiencing in the area of my solar plexus, the one that normally heralded the arrival of an ET contact. On this particular occasion I tried desperately hard to override the intrusion and not allow it to interfere with the work I was doing at the time.

I posted some mail and returned from the Post Office just after 5 p.m. I shut down the computers in the study and promised myself that I would relax over the weekend and not get involved in any new business. Annie and I had a comfortable evening watching a movie and, because of my workload over the previous couple of days, I retired quite early.

As is my normal practice, even though feeling tired already, I still read for a while before I turned off the light. The abnormal feeling in my chest was still there, but the idea of slumber was superseding any concerns I had about an interrupted night's sleep; this was probably wishful thinking more than anything else – sometimes one just wants to be left alone!

I must have drifted off at some point, because I don't recall putting my book down. I awoke in the night, reached out across the bed and found Annie silent and still, sleeping next to me. Lying there in the darkness I could see by the digital clock that it was just before 2.30 a.m. Then it began.

I was aware of something in the room. My eyes flicked about taking in the available details that were not completely hidden in the darkness. Just outside the bedroom door, on the ceiling in the hallway, there's a smoke alarm with a small green light that gives off a minute amount of illumination. From where I lay I stared at the area below the light. I should have been looking at a reflective white area that was the glossy surface of the cupboard door, something that would normally pick up all available light, but it appeared to be blocked by something. As I concentrated harder, my focus sharpened. I could see the outline of something. It didn't obscure the whole of the door; there were slim gaps at the side and a larger panel at the top to see. Thinking it was a shade of something familiar, maybe some exterior light was casting a shadow from the door of the bathroom or from Annie's study. I was about to close my eyes and go back to sleep. I must have imagined it, I reassured myself. Then the 'shadow' moved, but still I went ahead and closed my eyes.

With my eyes firmly shut tight, I found myself sinking, as if the bed's mattress was somehow giving way and sucking me in. No longer aware of my environment, I sank down and down, falling now with no apparent support for my body. Then I was still once more, lying on a firm surface that held me rigid from head to toe. Again I could feel my body, I breathed deeply, filling my lungs and feeling the air passing over my lips. My face felt cold. Even with my eyes still closed I could tell there was now a dim light around me. After what felt a long time, I persuaded myself to open my eyes.

It took a few moments but finally my eyes became accustomed to the light. At first it appeared hazy, I couldn't determine the depth or height of my surroundings. Lifting my head up and supporting my upper torso on my angled elbows I could now see all around me. Across from me, slightly obscured by my upturned feet, I could see the silhouettes of three figures. They stood silently with what looked like a table separating them and appeared to be roughly similar in height – relatively small that is, although I was still struggling to perceive certain aspects of size correctly.

I swung my legs down from where I lay and my feet skimmed the floor. Still straining to see the unidentifiable figures before me, I was relieved to identify one of them as Little Star. Her dark, straight hair hung loosely down around her pale face, but there was no mistaking who it was. A loose fitting garment, with sleeves that stopped half way down her arms and in length nearly reached her knees, covered her slender frame. Her arms were straight and out stretched in front of her with downturned hands placed flatly upon the surface of a table. The figures that stood either side of her were ETs – identical in pose and standing directly opposite each other. They wore garments similar to the one that Little Star had on, but the sleeves reached the hands and its length touched the floor. My sense was that I had not met these ETs before.

I shuffled towards them and stood within a couple of feet; they did not turn or respond in any way to my presence. Somehow I knew what I had to do. I stepped into the remaining gap around the table and stood facing Little Star. Her eyes were blank and unfocused; they appeared to look right through me. I continued to move intuitively and placed my hands palm down onto the table. The surface of the tabletop felt smooth and warm. I was immediately aware of a gentle tingling through the tips of my fingers that quickly spread across the surface of my hands. Remarkably I felt the area below my hands give way slightly, my palms sinking slightly into something that was no longer solid, but somehow gelatin in nature. I looked from left to right, into the faces of the ETs, but they showed no response to what was happening, although I could see that their hands were also beginning to respond to an apparent elasticity in the surface of the table. Once more, I looked to the eyes of my Little Star, normally animated with a sparkling life force, but now flat and emotionless.

The tingling sensation rose up my arms and I could feel a wave of heat beginning to move through my body, throbbing with a definite rhythm like an electrical pulse of energy. It filled me from the top of my head to the tips of my toes, engorging every fibre of my being in between, finally releasing itself and discharging from the ends of my fingers into the surface of the table. I imagined I would have seen sparks emanating from my hands if I had looked. Then there were thoughts in my head that were not my own. As the energy rose and started once more to build within me, I could sense my brain filling up with new information: a million images cascading over each other, fighting for a place of prominence in my conscious mind. At the same time I could sense a release of my own: feelings, emotions, thoughts, tastes, reactions, words, faces, sounds, places, smells – a rushing force of information being shared, it felt, with those who stood around me. I could feel the exchange of information as it pulsated from mind to mind like the insatiable throb of a relentless pounding organic machine.

How long this went on for, I am unable to say with any real understanding, for I became lost in its apparently never-ending rhythm. During that time, I communed with the minds of the three sentient beings that shared the experience – and they too communed with mine. We were as one, if that is an adequate description of what occurred. One thing I am positive of is that, based on my newly acquired ability to access previous memories connected to my ET experiences, I had never felt anything like this before and have not since.

"It is dangerous to let the public behind the scenes.
They are easily disillusioned and then they are angry with you,
for it was the illusion they loved."

W. Somerset Maugham

Food For Thought: Sheeple

A scarily large number of human beings on this planet are happy to be told what to think, what to believe and what to feel. Look at most (if not all) of organised belief systems; they are more often than not 'telling us' how to respond to what is happening in the world around us. Whatever happened to free thought, the ability to think for ourselves, to be discerning, to disagree without having to fight instead of debating? We have been made scared to do so, I believe. We have been made to seek and, in fact, demand to be told what to think. We have become sheep or should that read *sheeple*! Where is our sense of responsibility for our actions and for the words we speak? Those that have set themselves up as our 'conscience' have taken it away. Have you done something you shouldn't have? Oh well, sign on the dotted line and 'somebody' will forgive you. Garbage! We are being used and manipulated by whatever controlling bodies we ascribe to. Is the intelligence behind all of this control and manipulation coming from the same source though? If we disagree with what a 'visible' person is saying, then we are able to set a target to question that action, but if that force is unseen or hidden in some way, who do we turn to? Who responds to our questions? In fact, if the force is unseen and the instigator is hidden, do we even know something is being done to us? After all isn't it true that the most effective prison is the one without bars.

"If you tell the truth you don't have to remember anything."

Mark Twain

Experience: Nocturnal Contact in the Country

Home Counties, UK – Christmas 2005

It has to be said that I am not a huge fan of Christmas. I am not your proverbial Scrooge but I do not like the commercialisation of the whole event for one thing. Additionally, I don't much care for all the apparent (enforced) jollity surrounding the 25th and 26th December. I believe that I was different when I was a very young child, but certain events surrounding Christmas of 1969 polluted my ongoing appreciation of this religious and family holiday. I am not referring to an ET-related experience, but something that happened much closer to home and formed a debilitating belief system that went on to hamper and retard my emotional growth. So, when it comes to Christmas, please don't expect me to surface until the 27th because the two days prior are spent with the hatches firmly battened down. And don't even get me started about New Year's Eve!!! Please don't cry 'Humbug', I really would like to be different...honestly.

Anyway, on Christmas 2005 I responded positively to a heart-felt plea from Annie to attend a gathering of friends who had rented a palatial country residence for the holiday period. Begrudgingly I had to admit that everybody had a pleasant weekend, although on the last night I went to bed feeling a little rough, believing that I was coming down with some sort of influenza.

I awoke the next morning with a vivid memory that something had happened during the night with a distinct ET flavour. As is my ritual at these events I immediately focused inwards and activated my newly acquired skill and technique for retrieving memories that might otherwise wish to place themselves in a suppressed, compartmentalized area within my subconscious. I remembered going to bed, feeling achy all over and finding it hard to breath (because I was unwell, Annie was sleeping in another room, so as not to disturb me). I did not sleep well and tossed and turned for several hours. Then, at some point in the early hours of the morning, I was brought to full consciousness by a sound in the room. I sat up and saw, standing by the side of my bed, my wife Annie. Although a little alarmed to find somebody in my room in the middle of the night, the fact that it was Annie instantly calmed me. What suddenly quashed this feeling though was the sense that she was not alone. Standing just behind her and to the right was an ET: it stood just a few inches shorter than Annie (she is just over 5ft tall), a classic Grey in appearance who was clothed in a loose-fitting garment. From behind, the ET reached round my wife's body and grasped her by the wrist gently pulling her towards him. Annie turned, but not before she looked down at me and smiled. Her lips parted and they moved as if she was saying something but I was unable to hear the words spoken. Annie and the ET exited from the room via the open door and stepped into the dense darkness of the hallway. As is usual in this sort of situation, did I pursue? No, I lay down and went back to sleep! I know it's unbelievable.

Although I still felt remarkably ill in the morning – more intensely as I was not at home – I went to the bathroom, showered and sought out my darling wife, the best nurse in the world and just the person I need in attendance when feeling poorly and in desperate need of tender, loving care.

I told her of my nocturnal experience. Her eyes immediately defocused and she seemed to drop-out of the moment. Refocusing, she said, "I don't know, but that rings a bell, but only with regard to a vague dream memory. It's not clear."

"Well, stay with it, remember what I've told you about regaining a memory; focus in and try to retrieve it."

"Okay, I'll try."

The matter was not brought up again until we reached home and were sitting relaxing once more in the sumptuous surroundings of

Casa del Jones when Annie burst into animated activity.

"You said 'they' held me by the arm and pulled me out of the room?"

"Yes" I replied, not sure where she wanted to take this.

"Get the light and see if they left an imprint."

A couple of years before, Annie and I went to a BUFORA (British UFO Research Association) conference in North London. One particular presenter at the event spoke of certain Experiencers using a UV/Black Light to shine on their body after a contact experience to see if the ET had left any sort of residue that could be detected. He had confirmed that many people, including himself, had had success when carrying out this practice and that a fluorescent material was found under the light that was otherwise invisible to the naked eye. We had purchased a UV/Black light from an electrical store and had been waiting for an event to present itself to carry out this scientific experiment.

First of all I took off my clothes and Annie scanned my naked torso for a mark of some sort, but nothing at all was to be found. And then, as I was getting dressed, Annie placed the light down on the bed and its beam caught her arm, the area between her wrist and elbow. I could not but help expressing amazement: there, on her arm, were three distinct illuminated areas, all oblong and in clear 'finger-like' shapes and pattern. It looked exactly how one would expect it to look if a three-fingered hand had grasped your arm and left a bruise, but this time it was a highly illuminated and fluorescent bruise. I stopped dressing, picked up the light and shone it properly on Annie's arm. Her face was a picture: it fluctuated from amazement to fear and back again, through several other unidentifiable emotional responses.

We documented the event by taking several photographs of her arm. We were not to know at this stage but this was not going to be the only time when we would use the light after a contact experience and find some sort of ET residue left behind to confirm the visit. On other occasions I have had clearly defined circles around my nostrils; half moon shapes behind my ears, a straight line covering the scar on my forehead (that relates to the contact experience from my earlier encounters as a child), and marks on my feet, hands and knees. We were never to find any residue left on Annie again, although I shone the UV/Black light up and down her naked torso on many occasions – well, you have to carry out these scientific experiments at every opportunity don't you?!?!

"Perhaps being confronted with the facts of extraterrestrial contact shatters preconceived notions of our place in the universe. Resistance to new information is most intense when it rocks one's world view and if there is dissonance between new information and what we already believe, we either dismiss it as impossible or lash out at the messenger."

Unknown author

Experience: Kim's Website

Essex, UK – 2006

In 2006, I was approached by an old girlfriend of mine called Kim who had a business proposition for me to consider. Kim is a renowned artist with many of her paintings hanging in major galleries around the world. Her portfolio encompasses the wonder and beauty of natural history, as well as capturing the magical, yet elusive, portraiture of the human spirit.

She spoke of her desire to market a collection of her paintings as high quality prints and wanted to know if I would be interested in promoting her work on the Internet. Because of my background in marketing and sales and an extremely successful eBay business I once operated, she considered me the right man for the job. I told her that I would seriously consider the offer.

After mulling over the proposal, I agreed to giving it a trial. I said that I would come up with a selection of ideas and would arrange a meeting with her in due course. Kim already had a website and she gave me the password to enter it and to carry out alterations to the layout and copy. Within a week or two I had come up with a selection of marketing concepts and was well on the way to making changes to the website.

On one particular day I must have spent at least five to six hours entering the website, making an alteration and then exiting. A couple of days later I received a telephone call from Kim. She said she was very impressed with the amount of work I had done on the website and stated specifically how many hours I had spent actually making the modifications. I asked how she knew this. She replied that the software for the website has some very useful tools: one of which is to see who goes on to it, what they look at and how long they spend doing it. Strangely she said all this in a mock voice filled with sinister tones. What are you getting at, I asked, laughing at her suggested intrigue?

"On the day that you were on it, for nearly seven hours I think, I scrolled through all the data relevant for that day and I found some very interesting information."

"Go on."

"Well, there it was, the time that you entered and left, and what you did whilst you were there."

"Okay, I can understand that; it sounds natural to me, because I had the password and was making changes. Why all the creepy overtones?"

"As I said, I went through every single entry for that day and it was as if you had somebody following your every move."

"Explain."

"Each time you went in, the same person went in right behind you and when you looked at a specific part of the site, they looked too. Finally, when you left the site, so they left too."

"I can't believe it. What's that all about?" I thought for a second. "It's a shame we can't tell who it was."

"We can" Kim replied rather smugly.

"How?"

"Okay, part of the software enables me to see their user name and where they originate from... well, sort of anyway."

"What do you mean?"

"Well, it's not hacking, that's way beyond my capabilities anyway. All I'm doing is seeing what's recorded?"

"What did you find out?"

"I obviously couldn't find the name of the person, but I can tell you where the computer was that they were working from."

I didn't say anything; I wasn't sure I wanted to know.

"The computer," Kim went on, "is located somewhere on the network for Edwards Air Force Base in the United States of America."

"Are you suggesting that somebody on a military base in the U.S. is monitoring what I do when I'm on the Internet?"

"I can't say that they're definitely looking at everything you do, but by analyzing what you're up to through my site would suggest they have the ability to see what you're looking at in this instance."

"So it's not a civilian, but a member of the armed forces?"

"Well, it's specifically the Air Force, isn't it? Edwards Air Force base?"

"I suppose, by definition, you're right."

"Intrigued?"

"Worried more like," I replied uneasily.

My plans to work with Kim never went beyond this stage, not because of anything sinister, but rather that it just didn't work out. I don't know if there is anybody still looking at whatever I look at when I'm on the Internet. I don't know if they are able to read whatever I write when I'm on my computer, although I've heard that there is the technology to do so. I'm not sure if my phone is bugged but I have circumstantial evidence to suggest that it is. Also, I did have a spate of my post being opened before it was popped through my letterbox, but that could have been somebody at the Post Office messing around.

All in all, I don't really care. I'm not one of those people who have anything to hide and I am not going to make a stand over something like this regarding my civil liberties. If anybody is listening in or trying to find out what I know or what I'm thinking about, they only have to ask and I will share any thoughts and knowledge, or wisdom come to that, about absolutely anything. I don't have any secrets. In fact, I seek a platform to be able to share what I have experienced and what I believe it means. I believe the contact I am having with this remarkable intelligence is for everybody to share. What I am acquiring through the contact – the expansion of my mind and spirit – is to be spread to all four corners of the planet Earth; wherever there is a human soul who wants to know.

"Facts do not cease to exist because they are ignored."

Aldous Huxley

Experience: Shaven Leg

Central London – 2006

I have decided to include every little aspect of my contact experience, or what I consider to be associated with it in some way or other, even though at times it might sound inconsequential or even just plain old crazy.

And this is one of those examples. I suggest that you make up your own mind.

One day Annie and I went to visit friends of ours who live in central London, with the plan of staying over for the night. Their eldest daughter was temporarily shuffled out of her bedroom to the bedroom of her younger sibling, thus enabling us to have somewhere to sleep.

During the daytime I had had one of those 'peculiar feelings', the ones that I have gotten in the past that can sometimes herald a possible contact experience. I didn't mention it to Annie, as I thought it might make her feel uneasy, especially as we were in somebody else's home and obviously did not want anything to impact on them. So, I kept silent and hoped that I was imagining it.

We had a bite to eat and chatted to our friends for a while. We felt really quite tired so we ended up going to bed relatively early. As we began to fade away to sleep, I whispered: "I have a feeling that something is going to happen tonight."

I felt Annie shift in the bed and sit up. "Oh no, not here sweetie. Please not here."

I grimaced. "I'm sorry."

"What do you mean anyway, do you just have one of your feelings?"

"Umm."

"Maybe, it's just your imagination," Annie said, in a desperate attempt to reassure herself.

"Maybe" I replied, not believing my response at all but feeling too sleepy to explore the thought any further. I felt myself being drawn down, deeper and deeper. As the day had progressed I had become more and more sure that something was definitely going to happen.

We lay there silently under the sheets in our unfamiliar surroundings, then Annie finally turned off the light. She cuddled up underneath my arm and I felt her squeeze closer for security. I knew exactly what was going through her mind; but I don't think it was exactly the same for me. Finally, we must have both fallen to sleep.

Annie got up before me the next morning and went into the lounge to play with the children. After a while I went into the bathroom to commence my ablutions and found, to my utter astonishment, something on my body that had not been there the night before.

I immediately made my way back to the bedroom and called for Annie to join me. Within a moment my wife appeared. I stood there completely naked.

"Do you notice anything different?" I asked.

She looked me up and down and issued a gasp. "Oh my god, how did that happen?"

Annie's eyes were focused on an area of my right leg between my knee and hip, just fractionally to the right of centre. My legs are hairy, not too hairy, but hairy enough. In fact, they are hairy enough to easily notice when there is a perfect circle, about four inches in diameter apparently shaven into the growth of hair.

Annie didn't say anything; she just stood there and stared. I wasn't at all sure what she might be thinking. In fact, I didn't know what to think myself. The skin in the interior of the circle was completely smooth and its edges were exactly and precisely defined as if cut with a tiny lawn mower.

Annie finally spoke. "That wasn't there last night, when we went to bed, was it?"

"No."

"We would have noticed surely."

"Yes."

"How did it get there then?"

"I don't have the faintest idea."

Annie studied the area even closer. "What does it mean?" she said furrowing her brow.

I just shook my head.

"We must overcome the notion that we must be normal because it robs us of the chance to be extraordinary."

Steven Jones 2010

Experience: Just Point the Camera!

Home Counties, UK – Winter 2006

On another visit down to the exquisite Buckinghamshire countryside whilst we were house sitting for a friend, I had a remarkable and successful experience with my camera, taking pictures of 'invisible' things out of the bedroom window.

Let me explain: Annie and I had a bedroom overlooking the gravelled driveway in front of the house. To the left was another building, which was used as a garage and above this was a games room. We had done some minor research on the photographing of what had become known as 'Orbs'; small balls of light that appear after the photograph has been taken using an electronic, digital camera. Before the moment of capture, there appears to be nothing there, except apparently empty space. As there is little, if any, light pollution around the house, Annie and I thought we would experiment when we retired for the night.

I had my own technique: before I actually took the photo I would 'speak to the universe' and ask it to allow my camera to 'see' something and to be able to 'record' it. Pointing the camera out of the window, without really thinking about where it was directed, I repeatedly snapped away until I believed I had captured something. Reviewing my work in the small LCD window at the back of the camera we were absolutely amazed to see a whole spectrum of illuminated balls of varying sizes and distances showing up. There was a peculiar quality to the light of each ball and one could see a distinct rim encircling many of them, whilst others had detailed lines and patterns making up their surface. We carried on snapping away recording these elusive orbs, until finally they seemed to stop registering. Or maybe they just went away.

On the following night, I thought I would conduct a different experiment. This time I pointed the camera towards the night sky to see what might be recorded. Using the same technique of 'requesting' something to appear and to be captured by the camera, I nonchalantly directed the camera skyward without even looking, my face turned backwards at Annie as she stood behind me in the bedroom. I must have pressed the button on the camera ten times and then stopped. Reviewing all the pictures carefully, disappointedly we found absolutely nothing. We both had a sense that 'something' was out there, so I carried on. Snap, snap, snap. I must have taken about another ten again. But this time the results were different. Right there, in the dead centre of picture number seven was a UFO. If you can imagine two balls of light, connected with a bar of light, then you might be able to picture what the UFO looked like. It wasn't vague, it wasn't debatable, it was clear and pronounced, and it was most definitely a solid-looking object. The only thing was though, although we were not sure if it was there when I took the photograph – because neither of us was looking out of the window at the time – there was absolutely nothing there now. With great excitement, I began again. Snap, snap, snap. Once more, on a picture towards the end of my picture taking was an object; it wasn't just an unexplainable light, it was another UFO, but this time it was a very recognizable and classic shaped one.

I had to use the focus tool on the camera this time to bring the object clearer into view, but there it was, a classic 'saucer-shaped' disc. We couldn't believe our luck. We both wondered though, did we 'invite them' to appear for the camera or were they 'just passing' and we were fortunate enough to photograph them at just the right moment? I guess we'll never know. But it is nice to get these photos out now and again and relive the moment.

"Once upon a midnight dreary, while I pondered, weak and weary,
Over many a quaint and curious volume of forgotten lore,
While I nodded, nearly napping, suddenly there came a tapping,
As of someone gently rapping, rapping at my chamber door.
"'Tis some visitor," I muttered, "tapping at my chamber door.
Only this, and nothing more."

Edgar Allen Poe, *The Raven*

Experience: Solo in Laughlin

Laughlin, Nevada – February 2007

Towards the end of 2006 I experienced one of Annie's intuitive feelings regarding my attendance at the next UFO Congress Conference in Laughlin. It wasn't going to be possible for both Annie and I to travel there because my wife already had travel plans at that time to go to Venice with her son and youngest daughter.

I didn't really want to make the trip on my own, so I continued to grapple with this overwhelming sixth sense that I was supposed to attend. Annie insisted that I should listen to my intuition, so on a cold February morning in 2007 I was deposited at Heathrow Airport by my spouse and away I went to the warmer climate of the west coast of America.

Our good friends, Anne Cuvelier, Peter Robbins and Dr. Michael O'Connell, were all going to be at the conference so I was looking forward to teaming up with them to attend all of the presentations.

I won't go into a description of the conference or who the presenters were because this particular narration is about one specific evening whilst I was there and an unexplainable, but intriguing, event that occurred.

Wednesday at the UFO Congress Conference is 'Meet the Speakers' time in the evening after the day's presentations. The hall is restructured to enable tables and chairs to accommodate a sit-down 'nibbles and drinks' with all of the presenters from the whole week. They circulate from table to table taking direct questions from the audience and a 'jolly good time' is normally had by us all.

As it was only nibbles and drinks, I went back to my room to dress for the event via the Chinese food restaurant on the ground floor of the hotel. They offer a take-away service, so there I was struggling back up to the seventh floor with a scrumptious array of oriental delights for my evening meal. The plan was for the three of us to meet up later and slot onto a table with some of our other friends and acquaintances there that week.

I placed my delicious fare on the dining table in my room and turned on the television, hoping to find a movie to watch, before I returned back downstairs.

The next thing I knew it was 7 a.m. the next morning and my alarm was notifying me that it was time to rise for a new day. My Chinese dinner was still sitting there, completely untouched and now, obviously, stone cold. I still had on my clothes from the previous day. Sitting there rubbing my eyes, I couldn't work out what had happened; clearly I had missed the jollities of the night before, but I had no recollection of deciding to do so or even of feeling so tired that I must have just fallen asleep. I felt very confused. At that stage, I didn't feel anything else, but that was to change.

When I went down to breakfast Anne and Michael harangued me for not showing, but I had little defence except to explain the circumstances that I found myself in when I awoke. I speculated that I must have just lain down on the bed and fallen asleep; the only thing was that this behaviour was not like me, especially as I never, ever sleep through the night uninterrupted for one reason or another.

As I sat there during the morning's presentations I began to doodle on the pad that always sat on my lap, ready to take notes. What came out was a woman with big black eyes and long dark hair, parted in the middle and wrapped around her face. By lunchtime I clearly recalled somebody knocking on the bedroom door, just before I was going to start eating the Chinese meal the night before. When I opened it I saw a woman standing there, the same woman that I had drawn that morning. The strange thing was though that I remember thinking, at the time, that the woman looked a bit like Linda Cortile, my friend from New York, who was the subject of Budd Hopkins' book *Witnessed*. But my depiction, however, looked nothing really like Linda. And that is where my memory stopped again.

I spoke with Michael at lunch about this resurfacing memory and also showed him my drawing. He didn't make any reference to it looking like Linda, so my thought about her was obviously just connected to the woman that I believed had come to my door the night before. During the next break in the afternoon, Michael and I went into the concessions hall where a number of people can sell a wide variety of merchandise directly and indirectly associated with our subject matter. As I stood looking at one of the stalls selling DVDs Michael came rushing over to me with a book in his hand that he had temporarily picked up from the stall next door. The book in question was Budd Hopkins and Carol Rainey's book *Sight Unseen – Science, UFO Invisibility and Transgenic Beings*; it was already opened at page 252 and I could see a group of four drawings made by Experiencers of so called Transgenic Beings (a being with an amalgam of human and alien DNA and features.) The top two drawings bore a remarkable similarity to the one that I had done that morning and, what I found more amazing, was that one of them had been created by Linda Cortile.

Now, it might be suggested that I had seen this particular page in the book before, for it was true that I owned a copy myself, and I suppose that the drawing might have been imprinted on my subconscious mind. The fact that I had thought my hotel room visitor was Linda Cortile and that she had drawn one of the depictions of the Transgenic Being in the book was also rather intriguing. I must admit to have been somewhat taken aback by what Michael had presented to me and wasn't at all sure how to process the matter further.

I have never taken the incident further and as I have narrated it here is where it rests today. I often think about my visitor and wonder what happened next: did she come into my room, did I leave the room with her or was it all just a figment of my imagination?

"When the student is ready, the master appears."

Buddhist proverb

Experience: Flash Your Lights For Annie!

Carolina, USA – August 2007

One of our extremely dear and beloved friends is Mary McCormack who lives down in sultry South Carolina in the USA. She owns two properties, one of which is a beautiful, detached house nestled discretely in the forest where she resides most of the time, whilst the other is a luxurious apartment overlooking miles of golden beach and the dolphin-frequented ocean. As I may have already told you, our friends in America are all unbelievably friendly and generous in every way and our darling Mary is no exception. On this particular occasion she made her holiday home available to us for over two weeks. The location is most certainly one of those dreamed about places: with the beach literally just steps from the apartment and stretching for miles in either direction; the weather constantly and reliably warm/hot and dry (although we were there once when a hurricane stopped by for a visit, but it had made its way up the coast by the next day); and the people an embodiment of the famous Southern hospitality.

During our vacation there, we were visited by a mutual friend called Peggy, who lives in Charlotte, North Carolina. She is a leading light in the UFO community in the south and is much admired and respected by everybody who knows her. Over the short time we have known Peggy she has become a very special friend for more reasons than anybody will ever know. Annie and I invited Peggy and Mary over to dinner one evening and a wonderful time was spent talking about experiences and discussing plans for a group of support centres around the planet to support the lives of Experiencers, our ongoing contact with the ETs and our spiritual and intellectual growth.

Peggy told us that she believed the ETs are always very close by and that if one remains focused it is possible to communicate with them. I knew this to be true. She has a history of direct contact with them and assured me that the ones that she spoke with only ever had our best interests in mind. I asked her if 'they' would speak to me that evening if she asked them to or would they just do something so I was aware of their presence. I wasn't testing Peggy – why would I want to anyway – and I wasn't asking for a party trick because I have too much respect for Peggy, I just felt a need to do so... so I did. Just before she left she told me that 'they' had said they would do something before the night was out to make their presence felt.

Annie and I sat there talking after they both left (Peggy was staying at Mary's house): I sat curled up in a huge armchair (big enough for two), facing the balcony, the velvet black sky and the gently lapping ocean, whilst Annie was stretched out on the couch looking into the room away from the window.

Here is another wonderful example of not speaking up when something peculiar and amazing happens. As we sat there discussing our plans for the next day, two unbelievably bright lights burst like exploding fireworks out in the night sky. If I had held out an outstretched arm, I would imagine each light to have been about half the size of my clenched fist. They were not that high above the horizon, but high enough not to be a surface vessel. I don't think they were connected but they could have been. I suppose the distance between them – using the outstretched arm example again – would have been half a dozen clenched fists lined up next to each other. The burst of light began as a small pinpoint and rapidly grew, then hung there, suspended in the sky. Thinking of it now, it reminds me of a really large 'orb', a ball of light that has been captured using digital cameras. In the lights that appeared over the ocean that night there was an exterior halo that was brighter and denser than the rest of it and the interior appeared to bubble as if somehow alive. It must have lasted for at least eight to ten seconds, and all the time I did not react in any way or notify Annie about what was going on behind her back. And then it stopped. As if a switch was thrown somewhere, it just turned off.

I went back to my conversation with Annie, knowing in my heart that Peggy's ETs had kept their promise and made their presence felt. After a couple of minutes I said to Annie, flippantly in a way and with no sense of drama, "Peggy's ETs just flashed a couple of lights to me." I gestured out towards the window. "Out there above the ocean."

Can you imagine what my wife's response was? She went crazy; not believing it could have happened and she had missed it, and that I hadn't even told her when it happened. "Make them do it again," she pleaded and squealed in a petulant and hard-done-by child's voice. 'Please, ask them to do it again.'

Not really believing that anything would come of it, I closed my eyes and asked in my mind for 'them' to do it again. My voice pleaded in silent prayer. "Flash your lights for Annie. Please flash your lights again."

Annie had now turned around and was facing the window. She had opened up the glass, sliding doors to the balcony – enabling a clearer view, but letting out the cool, refrigerated air and also letting in the mosquitoes – and was now perched on the side of the couch, waiting and hoping.

I recited my affirmation over and over, with my eyes closed and my mind focused out towards the sky until Annie gave a shriek of delight! I opened my eyes to see two huge balls of light, bursting once more out in the black night sky. Annie was on her feet, jumping up and down and, like me, probably not really believing what she was seeing. What a night! What an experience! Thank you 'guys' for the amazing show you put on for us; it was very much appreciated (understatement of the year!).

" I decline to accept the end of man. I believe that man will not merely endure: he will prevail. He is immortal not because he alone among creatures has an inexhaustible voice but because he has a soul, a spirit capable of compassion and sacrifice and endurance."

William Faulkner

A Stimulus To Our Own Ends

The time had come to re-evaluate the role I play in my contact experience. It was clear to me that what was happening to countless numbers of diverse Experiencers all around the world went way beyond the surface picture as reported by them and speculated upon about by investigators of the subject. It meant something deeply profound, as far as I was concerned, and it is now our joint responsibility to determine that true meaning.

In Steven Spielberg's film from 1977, *Close Encounters of the Third Kind*, a seed is planted in the minds of the Contactees to identify an exact geographical location in the USA and travel to that point at a specific time and date and it is left up to them to find out how to proceed. I'm sure that the movie affected its viewers on many different levels, but now the message contained within its fictional narrative was starting to have an impact on the Contactees of today and we had become players in a real-life version of the movie.

I began to formulate a list of questions: What did the contact really mean to the individual participants and, more importantly, to the human race as a whole? On a personal level, how had my contact influenced my belief systems and how did it continue to shape my day-to-day behaviour? What constructive action could I now take, based on the growing awakening that I was experiencing?

I truly believe that from the moment we engage in the cycle of life to the time we die and then rejoin the long and eager queue to return to the material world, everything that we experience is recorded and stored forever more. Therefore, even when we have trouble recalling certain facts from our life it is not because they have been 'lost' but rather they are temporarily inaccessible. So, based on that belief I know that all of my contact experiences have affected how I have lived my life: sometimes in ways that are blatantly obvious and then in other ways that remain subtle and indistinguishable from normal thinking.

Elsewhere in this narrative I have spoken of my view that some memories continue to have an influence on my life without being there in the clear forefront of my conscious thoughts. I have referred to these types of memories as the 'wallpaper that hangs in the passageways of our mind'. As we go into the real life entrance of our homes, we rarely, if ever at all, identify and acknowledge the decoration that greets our conscious mind, but it has been proven that that initial impact stimulates our senses in an unconscious way. It is the same, I believe, with the contact experiences that are stored away in hard to reach or apparently lost 'memory files'.

I now have full and accurate recall of all of my ET experiences dating back to when I was just five years of age. They had always been there, I knew that, but some had been compartmentalized for my own safety. When my 'reawakening' began in 2001 a vital part of that process included unrestricted access to all of my contact memories. In doing so I began to understand certain traits that had made up my character over the years and had had a detrimental affect on my belief systems.

First of all, because I had incorrectly remembered some of the contact – especially in later years – as traumatic, I began to understand why I had been subconsciously fearful for most of my life. And it was because I believed I had no control over what was happening to me, in my ET contact and in life in general, but now I know that wasn't actually true, just a false perception. As you are aware by now I believe that the majority of contact experiences are the same, but are perceived differently based on the belief system of the individual. It would be rather naive of me to say that there are no negative experiences happening, because as with all aspects of life there is light and shade. However, based on my own experiences and research I feel I am able to make that statement. If you are a fear-based person it is very likely that your experience will be coloured by that viewpoint. There are a million and one perspectives that will ultimately effect an experience that occurs outside of the normal 'waking' reality: fear, religious beliefs, education, learnt behaviour, role models, parents, contemporaries, work place. So it is important to remember that the contact events are happening just outside the normal realm of experience, in a domain that one might describe as surreal, but not in the usual sense. Therefore, working out what is actually happening during a contact experience, recording that event accurately and then remembering it at a later date, can be distorted by an amalgam of sources, which lie active yet beyond our control in the workings of the mind. I have now been able to sidestep that normal human obstruction and know that we are all capable of doing so. I'm not saying that I still don't forget where I have put my keys or have a problem remembering certain items I need to purchase when walking around the supermarket. What I am saying is this: when I have an ET contact experience, it happens consciously and within the normal realm of reality; it is then stored accurately and is accessible whenever I so desire. But that wasn't always the case and in that way it had a crossover affect that infected my belief system like an undetected virus.

I don't believe that the ET contact is about the individuals who are instigating that communication. It is also not about the mode of transport that they use. Neither is it about where they hail from. It is also possible that the activities they carry out during their contact with us is only integral to their own objectives and not that important to our own. What they have to say may be the only element of the proceedings that we should take notice of. So if we remove all of the other components from the overall picture we are left with *us*. I firmly believe that the ETs are, therefore, a stimulus to our own ends: the spiritual evolution of the human race.

It is important to consider that the ETs are somehow connected to us, either through their point of origin and present home base, or possibly their genetic linkage to the human race. If we were to acknowledge these probabilities we can then see possible reasons for their contact and the content of what is being said.

The human race is on the brink of an upgrade: spiritually, physically, intellectually and emotionally. However, I do not believe that we will be able to achieve that goal based on our current software and hardware. Our bodies need to change on a physical level, so that we can function with the new information we will acquire to move forward. As I have said before this all sounds a little bit like the *Idiots Guide to Evolution*, but I do not have the ability to explain it otherwise and there is a need for me to speak in a manner that we can all understand; after all, what is the point of being so profound that no-one can understand what you are saying? Re-reading my words I believe and trust that I have managed to achieve this.

These alterations have been occurring for a number of years now and will continue up until, and probably beyond, a specified point of transition. The knowledge of this leap forward is recorded in the cultures of people from all around the world. The year 2012, I believe, is the focus for all of these events. 2012 will be the end of an era and the start of another. There will, of course, be certain events: earthquakes, floods, disruptive weather patterns, etc., as depicted in an ever-increasing number of books, films and in the media at this time, but the Earth is constantly making changes to its construction, it always has done so, it's a normal process. But please be aware that some catastrophes might not be natural and may be instigated and controlled by human hands. These human-manipulated events have already begun – can you identify which ones they are?

In the lead up to the 2012 time window there are interested and benevolent sources that will continue to play a hands-on part in assisting us in that change, but it is not all down to them. Through contact and communication the ETs are playing their part in stimulating us at this very important time of re-awakening. But it is not entirely their responsibility. The universe, as a conscious and sentient energy, is also a vital component of the bigger picture – clearly chapter two of the *Idiot's Guide to Evolution*! By and large, however, the most fundamental and significant ingredient for the success of this event will be what we ourselves actually contribute.

The human race has been issued with a clear invitation to the dance. But to gain entrance to the ballroom, however, it is strictly no jeans or sports shoes. A mirror has been held up for us to look in. Are we pleased with what we see? Some of us are dressed appropriately whilst others still require additional grooming. So how do we get our act together to ensure that we are not left at home whilst everybody else has fun at the dance? The answer is actually blindingly obvious and a no-brainer to achieve. The answer has been out there in the public domain for the longest time, since time immemorial, in fact. One word of advice though, if any reader has one bone of cynicism left in their bodies it is now time to have it extracted. You will not be able to hear the truth if you do not. Open your minds and lay aside your doubts. My complete skeleton used to be made up of pessimism, sarcasm, suspicion, disparagement, skepticism, distrust, doubt and scorn, but then I received my invitation. I can assure you that it wasn't easy leaving behind a way of thinking and believing that had apparently protected me all those years. After all, being a 'wise arse' is a great way of avoiding being a caring and sensitive human being. If I had only known in my younger years how easy it was to make friends and influence people I would not have taken the stony path to get where I am today. It was relatively late in the day that I made the changes, but at least I did find out that all I had to do was to be authentic, to be real, to be true. With that as a solid foundation, one can then move up to being caring; the next step is to be loving; whilst the ultimate goal is to be responsible for one's words and actions. I know we deal with a lot of challenges, after all life can be tough at times, but change, I believe, comes from the individual and not from the masses. Eventually, one becomes two, and all of a sudden we are of a like-mind. That's when the change occurs – when the individual grains of sand become the beach. We can achieve that by replacing one thought with another. It's not easy; I know that better than most. After surviving for so long by being a stranger to myself, it came as a major revelation – one of such empowering force – to be comfortable 'being in my own skin'.

Shakespeare wrote in *Hamlet*: 'This above all: to thine own self be true.' In following this advice I have now learnt to be kind to myself and in doing so, my life has become a place of safety where I no longer fear, because I know who I am and what I need to do at all times: which is to show love and care to everybody that I come into contact with, including myself, at all times, without exception. To take full and complete responsibility for every single word I say and every single action I carry out. If I can do that, so can you. These are not simply actions of the mind and then that's where it stops. Consider the theory put forward by the world of Quantum physics that there is an invisible 'field of energy' that connects everything in the universe, and that the actions of one thing or one person can affect everything else in existence because of that 'field'. If my words and actions become aware of the sensitivity and needs of others then their effect will have an impact not just on the people I come into contact with, but will have the potential to be of influence beyond my awareness. This is not airy-fairy, New Age waffle anymore, but facts and data being put forward by legitimate scientists in white coats.

By embracing this change and becoming aware, we gain entrance to the dance. By not listening and continuing to keep our heads buried in the sand, we become a version of Cinderella and stay at home. This is the opportunity that the human race has been waiting for. Don't forget though, there are agencies that would have us stay the way we are because it suits their purpose to keep us in a state of blindness and fear so that we stay forever in their power, beholden to them and propping up their inflated egos. Therefore, it won't necessarily be easy, but it is achievable.

I would like to propose a task, but it is important that it should be realistic. For such a long time I kept waiting for somebody to show me an idea of real worth, something practical that I could take out into the real world. Instead, all I heard were words, when what I desired was action. We all need a little push sometimes to ensure that we are moving off in the right direction. I have never wanted to be told what to do, but I do desire the occasional signpost. What I have to share with you might sound over simplistic, but understand this: it is only the key to unlock something much greater. What will come from you carrying out this action is the thing of wonder, I assure you. Please don't read on further with cynicism or you might just miss out on something extraordinary. There is an old adage about 'not running before you can walk', so please bear that in mind. Don't forget that at all times this is a 'do-able' thing. The completion of this task is a real, practicable step forward – not just a line in a book, a self-help book with good intentions but no practicable advice. It is a provable, constructive action to bring about change, change for you that is. Once again, it's not earth shattering in its content and will only reflect what I have already spoken about, but sometimes we all need to hear the starter's whistle.

So this is where you start. Consider the concept of being responsible for your words and actions when communicating and responding to everybody you come into contact with; do you think you could do that for a whole hour? In the workplace or in a social environment? Or even when you're completely on your own with just your own thoughts? Give it a try.

Then do it the next day. Remember, don't run. Then maybe do it the next day after that. Watch and experience the effect your kindness has, how the people respond to your care, to your stillness, to your listening. Then after each hour, see what it feels like not to do it and watch and experience the impact of not caring. It's all about mirrors remember, see what gets reflected back.

If you can complete a whole week, one hour a day for seven days, try it for two hours a day. Baby steps remember. Consider keeping a diary or just a notebook, so you can record your thoughts and feelings; it would definitely help, I guarantee it.

Consider these over-simplified examples: imagine going into a store and holding the door open for somebody behind you. Make eye contact, share a smile and acknowledge the thanks that you receive. Be in that moment. Hold on to that feeling for as long as possible and pass it on. Your grain of sand will soon become a beach. The effect of your one action will take on exponential energy if it is acknowledged in the moment and passed on. Take this one pure paradigm and translate it into a hundred different examples.

Your words and your actions will manifest something real. Watch and record the manifestation in yourself and then in others. My advice would be to not tell anybody what you are doing, then you will not feel self-conscious and in turn you will not be judged – either by others or by yourself.

I'm not suggesting that you become something that you are not. I don't want you to lose yourself. All I want you to do is to release something that is in there already. The more I think about it, the more I feel the need to emphasize the point that I'm not saying we should all become saints overnight: I want you all to remain the same people in essence, but just to turn 'your' page to the next section.

Then let the process expand naturally. Take it easy though, stay at a steady pace. The key to success will be making sure you do it everyday. Continuity and persistence is what it is all about. I am not suggesting you make the change of a lifetime overnight, but I am suggesting that you very carefully substitute one habit for another. To do this though you have to remain conscious during the process of carrying out this new action – you need to be aware of what you are saying and doing, and not operate on automatic pilot like we normally do when attempting to stumble our way through everyday from waking to sleeping.

So, give it a go, would you? It would be an interesting exercise if nothing else. Remember, 'This above all: to thine own self be true,' because if you cheat or don't keep up with the discipline of this task, try looking in the mirror and telling yourself what has happened. If you can achieve it you will be making the best friend you have ever had – you! (Sorry if that sounds cheesy, but it really is so true – and that is from the most cynical man in the world prior to the year 2001.)

Seriously, this really does mean something important. In the next few years, the human race will take part in a huge shift in its consciousness and if we are ill-equipped to do so, the process will be quite unpleasant. Look at it like this (and once again, forgive the cheesiness); the planet Earth is pregnant, it really is *Mother Earth,* and the spiritual evolution of the human race is the baby within. To ensure a healthy and comfortable birth, and to make sure we emerge capable of residing in the upgraded environment that we will find ourselves in, we need to be capable of operating in a new vibrational rate, one that is not as dense as the one we are currently in. Our new Earth will be a very different place and if we are unable to make the necessary alterations to all aspects of our being, we will not survive.

By carrying out the exercise as I suggest, it will begin to have an impact upon the physical and that will ultimately manifest in many ways, with the most important being a physical level, directly altering and upgrading aspects of our DNA make-up that will remain unseen at this time. Continuing to embrace the Quantum theory of a 'connecting field', the changes that will occur to you will spread out once more and have a positive effect within the 'field', and don't forget that the positive aspect of change that is happening to others will have a compound affect on you: 'Then one will become two and all of a sudden we are of a like-mind. That's when the change occurs.' The DNA change then becomes a permanent change and therefore our children and grandchildren will inherit the Transformation Years and pass it on. This is not science fiction, but science fact.

There are moments in the lives of each and every one of us that stand out. They are normally very easy to remember, although they may not always be pleasant memories. They are the events that are spoken of when asked to remember an occasion of importance, a time that leaves a profound imprint on our psyche.

I have three such memories.

On Friday 31st July, 1987 my son was born. He arrived in the early hours of the morning. There was one moment during his journey where we all became a little concerned, but after the professional assistance of the nursing staff the moment soon passed. My son was taken away for about 30 minutes and then brought back wearing a tiny woolen hat to *'keep his liccle head warm'* we were told. A robust and curvaceous Caribbean midwife handed my son over to me to hold. And then it happened.

I cradled my son in my arms and looked down into his eyes. I welcomed him into the world and then I began to cry. The tears fell from my eyes onto his face. I was overwhelmed with love, the purest love I had ever felt.

As I was driving home later in the day, a song came on the radio by a young British singer called Alison Moyet. I believe the universe was talking to me at that moment. As Ms. Moyet came to the chorus she sang, 'I go weak, weak in the presence of beauty'. For the second time that day I began to cry.

My son's arrival denoted an important transition in my life and his continued presence in my world remains just as significant and vital. He is and always has been a 'gift'.

I called my son, Reece Joseph.

In 1991, my dear friend Jacqui knocked on the door. I was still married to Reece's mum at the time and we all lived in east London. Jacqui told me of an advertisement that she had seen in the local newspaper, which told of a 'Psychic Development Class' to be held every Tuesday night not far from us. As we were both extremely interested in the subject matter, we decided to sign up.

So, on a Tuesday evening a couple of weeks later, Jacqui and I parked outside the address that we had scribbled on a piece of paper. As I got out of the car I looked up to the front door to check that we had the correct number. Just by the steps, which led up to the house, was a For-Sale sign. In seeing the sign I felt a profound sense of unease and then I heard a voice in my head, which clearly stated in an unforgettable voice, "You've finally found her again and now she's going to move."

I stopped in my tracks with the words still softly ringing in my head. Jacqui enquired if I was all right and encouraged me to get a move on.

I rang the bell and waited for it to be answered. When the door was finally opened, I got another shock: without having ever met Anne Ashley before, the tutor of the course, I totally and absolutely recognised her.

For the next few years, I wrestled with this dichotomy. During this period I also fell head over heels in love with Anne, but she never knew anything about my feelings. Due to problems relating to my own inability to make the marriage work, and definitely nothing to do with my feelings for Anne (which remained hidden), Reece's mum and I finally parted. The first few years after our break up were extremely painful and difficult for all of us, but we ultimately found a light at the end of the tunnel. I finally found the courage to tell Anne about my feelings for her. We then fell in love together. Anne became Annie and eventually we got married.

I don't know if I ever thanked Jacqui for responding to Annie's advertisement, the one that enabled me to follow that trail of golden breadcrumbs to her door, but without dear Jacqui passing on the information where would I be today? It's too scary to think about! So, thank you Jacqui – bless your heart!

Time passed. Life happened. Annie was always Annie, whilst I continued to search within for who I really was.

And then on Sunday 25th May, 2008, *I found myself.*

Up until that moment the universe had placed numerous obstructions in my way, which were designed to make me stop and look at my life, to decide if I was happy with how things were or would I like to make a change. On these occasions, small alterations were made; fine-tuning took place, but somehow not enough to make that almighty shift that my life deserved. Then the universe got fed up with waiting and placed an obstruction that I could not get round. So finally something wonderful happened. The transition was painful beyond belief, but I can only suppose that at the end I experienced something along the lines of how I imagine a new mother might feel after giving birth.

The details of what happened in this pivotal moment in my life are so intensely personal – more so than anything else that has ever happened for me – that I literally cannot find the words to properly or adequately describe what occurred. Even if I could, I'm not really sure if it would serve a constructive purpose. It is my belief that all that needs to be said is this: from that moment on *I began to take responsibility for my words and actions.*

Those are my three memories. In a way I believe they are all intrinsically connected.

I would like to add a fourth: *This* moment; *this exact moment in time* when I wrote this section and the feeling I experienced when I truly believed and embraced the concept of you reading what I have written and, by doing so, making a change within yourself.

"A sufficiently advanced technology is indistinguishable from magic."

Arthur C. Clarke

Food For Thought: The Golden Breadcrumbs

I have recently discovered Noetic Science and the further study of this, I believe, could be instrumental in helping us all to understand and appreciate what we've been talking about so far. Therefore I will take this opportunity to share some specific points on the subject that might work like 'golden breadcrumbs' to take us all to a different place in our awareness and perspective of the world in which we live.

The Institute of Noetic Sciences (IONS) researches human consciousness and its potential. The word Noetic comes from the Greek word for 'inner knowing'. The Institute was founded in 1973 by the American astronaut Edgar Mitchell and has its headquarters in San Francisco. Further information about the work they do can be found through their website (www.ions.org), and their research currently being undertaken on 'consciousness studies' is covered in their quarterly magazine, *Shift: At The Frontiers of Consciousness.*

Einstein noticed through his work that small particles of matter could 'communicate' with each other, no matter the distance between them. He was unable to understand how this worked and considered it to be 'spooky action at a distance'. Since that time modern science has continued to work in this area of study, one of which is the Joint Quantum Institute and University of Maryland Physics Department, which now believe particles communicate with each other by using 'quantum teleportation,' where a unit of information is communicated from one particle to another over any distance, which has a great bearing on the concept that our thoughts have a very real and tangible affect on the minds of other people and the world that we live in.

Additional work carried out by the Global Consciousness Project, based in Princeton (http://noosphere.princeton.edu/intro_bottom. html) found that, in particular, 'group consciousness' can have a profound effect when used in conjunction with the study of Random Event Generators (REGs) in various locations throughout the world to see if major planetary events made an impact on the REGs. A specific test that was carried out was produced by 200 'coin flips per second' being generated (the 'coin flips' corresponded to 'ones' and 'zeros' rather than 'heads' and 'tails'). Usually the results would produce quite random results without any specific pattern, but the scientists noticed that when something 'big' happened on the world stage the 'coin flips' stopped being random and a sequence of 'ordered numbers' were recorded. An example of this happened on 11th September 2001 at 5.30 a.m. (Eastern Standard Time) when the largest 'spikes' were recorded for a half-hour period and then sporadic unexplainable anomalies (i.e. patterns) in the hours leading up to the attacks on the World Trade Center Towers in New York City. This is but one of many unexplainable examples that have confirmed the work of the scientific community that is continuing to study Noetics. Through this it would appear that the 'mass global consciousness' can foretell major events; that 'thoughts' have an effect on the 'physical world', that our consciousness has a very real and tangible consequence and that there is truth in the saying 'mind over matter'.

Very interestingly, in the New Testament the Gospel according to John puts great emphasis on the Word as the spark that triggered Creation. John 1:1-3 states: 'In the beginning was the Word; and the Word was one and the Word was with God.' The 'Word' as expressed here is the logos or the creative word. Therefore, might it be understood that 'the Word' – or more correctly the *intention* behind the Word – is what created our reality. It's a thought...

"Do not speak unless it improves the silence."

New England proverb

Experience: Can You See Us Now?

Home Counties, Central London and Suburbs – February 2009

Annie and I have a relatively uncomplicated life and our needs are very simple. Our most enjoyable pastime is just being together; be it talking or debating, watching a movie, going to the theatre or travelling, romantic escapades – being in each other's company provides a wealth of fulfillment, satisfaction and stimulation. When we are apart, which is not very often, Annie likes to socialise with a small group of female friends – some friendships dating back over 50 years – whilst I spend my time writing, reading or in our art studio.

When I do venture out I love the mobility and freedom of driving, all the time either listening to music, talk radio or audio books on my beloved iPod. On one particular Friday in February 2009, I was en route from Cambridgeshire back across the county border into Essex when I was quite taken aback by the beauty of the countryside; so much so that I actually stopped my vehicle and pulled over to the side of a country lane. I sat on the brow of a hill with an unrestricted view across the sweeping and rolling landscape. The clear sky was a crisp ice blue, dotted with birds that hovered above the ploughed fields occasionally descending at speed towards unseen prey. Through the windscreen of the van it appeared to me like a lush, vibrant, oil painting within a contemporary frame; the moment magically captured forever by the mastery of the artist that is Mother Nature. I felt such a sense of wonder rise within my chest that I breathed deeply, taking in a long rush of invigorating country air, filled with an abundance of earthy aromas.

Suddenly there was a flash of light, just as if a photographer's flash bulb had exploded without warning, my vision momentarily blurred by a blinding white light. As my vision instantly returned I was astounded by what I saw, it was just like the high definition picture on my TV. Every colour had increased in luminosity tenfold; each line and edge was now more sharpened and defined; the depth of the picture had taken on additional three-dimensional qualities that simply leapt out and assaulted my eyes.

And then there was a voice. Without really thinking, I turned my head towards the satellite navigation system. The voice was so real, so genuinely authentic and right there in the cab with me, that I truly believed it was being emitted from the small black box that was perched before me on the dashboard. Obviously it was not. And then it spoke again, the very same words that it had spoken before: "Can you can see us now?" The voice was delicate, feminine, inquisitive, and full of depth. The words were not resonating in my head, they were being heard, most definitely heard, by my ears; they were being spoken close by.

I realised I was no longer breathing and quickly gulped in a breath. My eyes darted about the cabin and beyond. The mellifluous inquiry hung suspended in the air awaiting an answer. I don't believe I had ever been more present in the moment.

Then, the question was asked once more and for the final time: "Can you see us now?"

My response was a whisper, barely audible: "Yes."

Then another flash of light. Once more, I was momentarily blinded, my vision returning in an instant. All was the same once more: gone was my sublime, awe inspiring, high-definition vision of the surrounding countryside and in its place returned the normal visual interpretation of my world. In truth, it remained breathtakingly beautiful, but would now forever be somehow inadequate.

I don't actually remember starting off again. I wasn't missing any time and remained strictly back on course; but I stayed in a 'twilight' moment for the rest of the day, with that melodious question still turning softly, over and over, in my mind: "Can you see us now?"

Eleven days later while shopping for an hard-to-find food

ingredient, but this time in the inner city, the sublime essence reached out and touched my life once more. I was driving through an area called Dagenham in the east of London, when I saw through the windscreen of the car, a helicopter circling quite low in the sky. The next time I stopped to get out of the car I noticed that it was still there but now it was pretty much directly above me. I stood for a moment trying to see any identifying markings. As far as I could make out the colour of the metal work was black with no discernable identification of any sort. I thought this odd as I believe that any airplane or helicopter, due to some sort of aeronautical regulation, has to carry something to enable identification. I continued on, all the while keeping an eye on the helicopter and attempting to hear the whop-whop of its rotor blades through my open window. After driving for another ten minutes, I alighted once more to find it still hovering exactly over my position although noticeably lower; so low in fact that pedestrians were stopping to take note of this aircraft flying so close to the ground. Due to its altered altitude I was able to see that the windows were actually blackened out.

I decided to telephone Annie and let her know what was going on, but I couldn't get a signal, so, once more, I continued on my way. As I drove, I was still maintaining the thought that this was all purely coincidental and that I was witnessing an event that had no direct connection with me, but was relevant only to my present location. However, I very soon let go of that theory and embraced a new one.

Because there right above my head and close enough for me to hit it with a stone was the helicopter once more, hovering with its black shark-like nose turned down as if it was 'looking' directly at me. At that moment I accepted that the helicopter, my presence and geographical location were all unavoidably connected. Was I being followed? Could this be an act of intimidation? I really wasn't sure what was going on. However, when I tried Annie on the phone once more I finally got a signal and my text message went through. At the exact moment when my screen stated 'message sent' the menacing hulk of the large black helicopter that had been my travelling companion for the last half an hour, rose up, turned acutely and took off at a speed I didn't realise was possible for a helicopter. Coincidence? Maybe. But who knows?

The whole incident played on my mind for the rest of the day. About an hour or so later Annie rang me responding to my previous text. Strangely enough, she had only just received my message. By the tone of her voice she was clearly concerned about the intimidating nature of the event and wanted reassurance that I was safe. I assured her I was absolutely fine, because I *was* absolutely fine. I have always said that if there are any 'human agencies' out there that want to know, in the finest of details, about my contact experiences, they only need to ask me and I'll tell them everything I know.

One might imagine that the day could not have reached any new levels of surrealism, but you would be wrong. I pulled up outside the home of a friend of mine aiming to deliver a parcel. It was a solitary country house on an isolated lane on the border of London and Essex. Still absentmindedly checking the sky for new arrivals, I reached into the trunk, turned and bumped straight into a person who was standing right behind me.

Now, over the next sixty seconds, under normal circumstances, there should have been alarm bells ringing in my head to indicate that something was definitely not how it should be. Either the bells were broken or I just didn't hear them. Because of the house's isolated position and the fact that I hadn't pulled up right outside its drive but on the other side of the lane adjacent to an open field, I can't really say where this 'person' appeared from. The 'person' who confronted me was extremely tall, much taller than I am in fact, and probably taller than a human being has the right to be. I can't even remember what she looked like except that I am positive that she was 'female'.

She said, in a voice that was delicate, feminine, inquisitive, and full of depth, "Still looking for helicopters are you?"

"Yes," I said with an embarrassed smile, not questioning how she knew what had previously happened.

I felt like an intimidated child.

"She...needs to know that they can't cause you any harm," the 'female' said in a reassuring purr.

"I'll let her know. Thank you for that." Who was 'she' and what did that mean? Where were those alarm bells?

Leaning towards me and over me, this towering presence reached out an arm and with the finger tips from three unusually elongated digits, pronged my forehead with unexpected force, sending it back slightly with a shudder.

"You can see us now can't you? It's time for you to wake up and remember!"

And then she wasn't there anymore; but I didn't register that for about ten minutes. For no apparent reason I placed the parcel back in the car and drove away.

The M25 motorway that took me a few miles to the next junction with the M11, which eventually led me back home, was only about three miles down the road from that last stop. Surprisingly, it was only as I was about to filter off the roundabout onto the slip road that led onto the busy flow of motorway traffic, did I question: Who was that woman? What did she look like? How did she know the things she did? Where did she come from and where the hell did she go?

And guess what, I still have no answers to any of those questions.

Of course, I can speculate. I have a feeling that the 'female' was a Light Being, an ET of some description, maybe even an alien/human hybrid; or that her appearance may have been some sort of 'screen memory', projected into my extremely limited and easily manipulated field of vision to enable acceptance on my part – but would I have then 'felt' her touch? I think not. As for her appearance I have no crisp and defined recollection of what she looked like, except that she appeared to me as an unusually tall, human-looking woman. With regards to her awareness of the situation I can only assume that she had knowledge of the preceding hour because 'they' are in constant communication with me and, in a way, are my guardian angels in my limiting human interpretation of things – and this was shown in 'her' attempt to ease Annie's concern by saying "she... needs to know that they can't cause you any harm" in reference to the intimidating presence of the black helicopters and their occupants; and finally, where did the 'female' come from and where did 'she' go, I would hazard a guess and say that it was *behind the scenery*!

That evening after the encounter with the 'female' Annie got out the UV Black light, which I have mentioned before, and there they were: three illuminated shapes on the middle of my forehead, corresponding to the marks one would expect to be left by the tips of three overly large fingers. However, the marks from the fingertips were in a straight line, all right next to each other. Why is that strange, you might ask? Well, lean forward and tap someone of the forehead with the tips of your three middle fingers. Because the centre finger is longer and hits the skin first it will naturally push up and form the top of a triangle with the other two fingertips below and to either side. It's just awkward and difficult to form a dead straight line. Therefore, the inference is that the 'female' had fingers all exactly the same length, which is peculiar, yet has been reported by Experiencers when encountering human-like ETs.

"Anyone who keeps the ability to see beauty never grows old."

Franz Kafka

Experience: Sixty Candles

UFO Congress Conference, Laughlin, Nevada — February 2009

Unknown to us at the time this was to be our final attendance at the conference. In 2010 it was be announced that the administration and production of events would no longer be the responsibility of the Brown family and their team, and because of financial restraints we would not be there to hear these sad words and thus unable to attend the conference's swan song.

But in 2009, all of this was probably undecided and unknown to the masses, so our attendance would not be spoilt by the news of the approaching storm. In fact, we were there not only to enjoy the conference but to celebrate my darling wife's 60th birthday.

On Thursday 26th February, after the final presentation of the day, a group of us, including Annie and myself, Rudy Schild and his wife Jane, Anne Cuvelier and Barbara Lamb came together al fresco at a restaurant in the Flamingo Hotel. A jolly time was had by all, only to be crowned by a collection of beautiful gifts and a roaring chorus of *Happy Birthday To You*, led by the dulcet tones of Jane, who is a wonderfully accomplished and successful opera singer.

Upon waking the next morning I was suddenly hit by the memory of a surreal experience that had happened during the night. Rolling over on the pillow to look at Annie, I said, "Do you remember what happened last night?"

She hesitated for a moment, went to say something and then stopped.

"Anything?"

"No...not *really*. Maybe."

She didn't seem sure.

"Do you want me to tell you?"

"Go ahead."

I waited a second or two, collecting my thoughts and then began.

"At some point during the night, something woke me up. The room was lit up. It wasn't from outside because the curtains were still drawn and it wasn't the lamps inside the room. I didn't know what it was, but I could still see.'

Although Annie had clearly just woken up, I now had her undivided attention.

"You were sitting up in bed right next to me, on top of the covers with your legs tucked half-way beneath you and turned to your left."

I looked into her eyes for recognition, but none came as yet.

"Perched on the bed, right in front of you and looking your way, in exactly the same pose as you were, but with legs crossed in a full lotus position, was a petite and frail-looking child."

Annie caught her breath. I closed my eyes, recreating the picture in my mind so as to describe it to the best of my ability.

"I'm sure it was a little boy, probably about five years old. His hair was blond, very fine and parted in the middle. It was pushed behind a tiny pair of ears and appeared to be of shoulder length. He had the most amazing pair of crystal blue eyes; that could've been the source of the light in the room, for all I knew, considering how bright they were!" I thought for a second. "Actually I wouldn't say he was frail... *delicate* would be a better way to describe him. His features were like a cherub and his skin, which was very pale but not unhealthy looking, was smoother than anything I'd ever seen. His upper torso was just slightly tilted towards you. In my mind, I saw his clothes as pyjamas; they were silky-looking and had an opened jacket-top and trousers. His little feet were bare and reminded me of a pair of newborn baby's feet, not in size, but just in texture." I stopped and opened my eyes. "Anything yet?" I asked Annie.

She smiled softly, but didn't reply. I closed my eyes once more and carried on.

"He had tiny little hands that made me think of a bird's claws.

His fingers were interlaced and sat softly in his lap. You were staring at each other. Really intently. Staring. I looked closer at you and it seemed as if – from my side view – your eyes were shining too. In fact, taking you both in at the same time, it was as if you were both sharing a bubble of light."

I took in a deep breath and opened my eyes again. Annie had a blissful look on her face. I imagined she was reliving the moment or just picturing what I was saying. After all, this could've just been a dream on my part.

"You leant towards him and your foreheads were nearly touching. Both of your mouths were moving, but I couldn't hear any sound. Every now and again, one of you started to laugh or that's what it looked like. You seemed totally oblivious to my presence."

I then remembered something. "I tried to interject as you were speaking to each other, but every time I did so the 'little boy' put a finger to his lips in a 'schoosh' gesture, indicating to me – although he didn't look directly my way – to be quiet and not interrupt. I attempted this a few times, but he just kept laughing and schooshing me! You really seemed to enjoy that part of it."

"This is all starting to resonate with me. Although I can't remember exactly what you're saying, I am definitely getting a 'sense' of it," Annie said.

I then told Annie that I recalled her face becoming very serious and it looked like she asked the 'little boy' something, whilst gesturing back to me with her right hand. Thereupon he nodded and put up one of his tiny little fingers. Without turning in my direction I remember hearing Annie's voice saying to me, "I can't tell you everything we're talking about because it's a secret. But I can tell you just one important thing."

I didn't answer, but remember sitting there waiting to hear what it was.

In a low voice, whilst continuing to stare directly at the 'little boy', Annie said, "You mustn't worry about the hole. Everything is exactly how it should be." That was it. Nothing more. I mustn't worry about 'the hole'.

I lay silently in the bed waiting for Annie's reaction to what I had just told her. I opened my eyes, but hers were now closed. Then she stated, "There's more, isn't there? It didn't just end there did it?"

I thought for a moment, taking myself back to the picture in my memory. "No. The 'little boy' and you embraced, you had a really nice cuddle. Then, in a second, the light was gone and so was the 'little boy'. I watched you lie back down. I think...we both just went to sleep."

When Annie and I discussed the experience later on with our friends at the conference, it was suggested that 'the hole' reference might relate to something Rudy had been talking about in his presentation: Black Holes and their relativity to the spiritual nature of Human Beings and the ET connection. Somebody else proposed that it might be about the Ozone layer around the planet and that the 'supposed hole' in it might not necessarily be a bad thing and that it could actually have come about to disperse the dangerous atmospheric poisons building up in the air.

My sense of what I had experienced – I did not believe it to have been a dream – was that there was a very special kind of relationship between Annie and the child and that 'he' had paid a visit because it was her birthday. I believe that the child was an ET.

As we all sat there digesting and ruminating upon my experience, one last and very important memory came back for me.

'*He called me something.*'

"What do you mean?" everybody asked.

"It was either as he was leaving or it might have been when he told Annie about 'the hole', but he referred to me as... it sounded something like 'eti-en'."

A moment or two passed in silence.

"Could it have been Etienne?" somebody asked.

"Why, what's that?"

"Well, St. Etienne is a city in France. It was named after... Saint Stephen."

"Interesting!" everybody chorused.

"Another thing I've just thought of," Annie added. "Maybe it's a little joke, a play on words."

"Go on," I said.

"The first two letters of the word... *ET*–ienne!"

So they do have a sense of humour after all. I always knew they did.

*"Leaders are visionaries with a poorly developed sense of fear
and no concept of the odds against them.
They make the impossible happen."*

Dr. Robert Jarvik

Social Event: First Presentation at the UFO Academy

London suburbs – May 2009

In the spring of 2009 I was invited to speak at a new conference event in Watford, a suburb northwest of London. The invitation came from two empowered and courageous individuals – Catrine O'Neil and Harry Challenger – who truly believed that there was 'a hole in the market'. I happened to agree with them wholeheartedly.

Compared with America this type of event is few and far between, especially ones of substance and worth. There are still too many congregations of people gathering who remain happy and content to regurgitate tired and pointless stories of unidentifiable lights in the sky. As far as I am concerned – as well as the progressive wave of people who exists throughout the world – I feel we established the fundamentals of this subject matter many years ago. Therefore, I was extremely pleased to discover that this event was free from the deadwood of yesteryear with regards to the 'lights in the sky'.

As we enter into September of 2010 the Academy is about to present its third gathering and it continues to grow in both enthusiasm and 'bums on seats'.

If you will indulge me, I would like to share with you just a brief excerpt from my presentation on that day, which does not include reference to any specific experience but reflects upon my thoughts and feelings as I brought my talk to a close:

'UFOs are real. By that I mean they are physical, nuts and bolts modes of transport that are flying around in our skies – and to most they remain unidentifiable. There are real people inside these vehicles. That is to say they contain pilots that fly these craft. The origin of these craft and their occupants are sometimes from a very, very human source, whilst others stem from somewhere more exotic. Occasionally these pilots land their craft and make contact with human beings. What gives me the right to make these statements? For the best part of 50 years I have had an up front and personal relationship with the non-human occupants of some of these craft. I have seen these so-called UFOs close up and I have been inside them too. What shall we call the occupants of these vehicles – ETs, Aliens, Space Beings? Ultimately they are people. They might sometimes look and behave very differently from us, but in my experience they are all people. On this planet, how more alien in appearance can you get than a forest-dwelling pygmy to a tall blonde Swedish city person? An Australian aboriginal to a classic European Caucasian? An Inuit (Eskimo) to an African American? The comparisons are endless – but we are all just human beings. And it's the same with the so-called ET Aliens.

'There are many things to be got from our contact, but I have found that the most important aspect is the opportunity for us – the human race – to look at ourselves and see how we live on this planet and interact with each other. What am I? An Abductee? An Experiencer? Maybe even a Participator? Guess what, I'm a person too. My label is not important, but my experiences are or, more importantly, what they mean. Do I resent my contact experiences? Am I afraid of them? I used to be – at times I believed I was terrified – but not anymore. I believe that all of our contact experiences are pretty much the same, but are ultimately coloured by our own individual belief system. I used to be a scared little rabbit caught in the headlights of life, therefore my contact experiences were

scary. I am no longer scared of life, so my contact experiences have changed accordingly. So, that's it – UFOs are real and there are real people inside of them. After more than 60 years of undeniable proof and evidence to support this statement, isn't it about time that we stop faffing about debating this simple fact and move on? We don't need the support of our governments! We don't need confirmation from our scientists! We don't need disclosure from the powers that be! We are our own empowerment! We're not children – it's time to grow up. It's time we became personally responsible for this phenomenon and what it really means to all of us. For me my experiential contact is all about choice, responsibility, care and, most importantly, discovering the true meaning of love. For those who know me, they would be surprised, if not shocked, to hear this cynical old bugger say that love is all-important. Going back over two thousand years to a beautiful visionary young rabbi, who said that love was all important, to the 20th century, where a young Liverpudlian singer told us all that 'Love is all you need'. And through my contact with these remarkable alien intelligences, I too have discovered that the very simple answer to all of our problems is that, truly... Love Is All We Need. It's the answer to absolutely everything.

'My first conscious memory happened when I was a very young child. I believe that when we arrive in this physical life we are all handed a box of toys. These toys are up and above our normal five senses, these toys are extra-sensory. They are here to enhance our lives. However, as we progress in life and become part of the system, our toys become dull and are more often than not completely removed from our toy boxes. When my first contact occurred I know that all of my extra-sensory toys were still in the box, but they began to disappear one by one. Over the years I have finally got back my full set of toys. I am working towards everybody understanding what is happening to all of us and the opportunity we are being presented with. But it's vital that we speak the same language. I have discovered that life, after

all, is very simple – so let's keep the discussion very simple. Let's not unnecessarily complicate it just to appear clever or hierarchical. Some old Greek geezer called Pythagoras was the first man to call himself a philosopher. Before that time the wise men had called themselves 'sages', which was interpreted to mean 'those that know'. Pythagoras was more modest. He actually coined the word philosopher, which he defined as 'one who is attempting to find out'. I consider that a much healthier position to be in. Therefore, by definition, we are all philosophers, would you not agree? In my continuing journey, I haven't really come across anybody yet who knows what is really going on, but I have been blessed by coming into contact with an array of enlightened individuals who are 'attempting to find out'. We are very fortunate to be alive at this time. This is an extremely auspicious time to be a human being on the planet Earth. The eyes of the universe are on us and what will happen here over the next few years. We are the greatest show in town! The planet Earth is very unique in so many ways. We are one of the very few planets that possess free will. We as human beings have the choice – every single moment of the day – of what happens next in our lives. The planet Earth is so beautiful and exclusive in its diversity of life forms. Nowhere else in the universe will you find this type of multiplicity. The human race has such great potential for beauty – we are able to express such profound love for each other; such incredible acts of humanity and goodwill towards each other. In the darkest of times we see the human race rise above all that stands before us to overcome the greatest of challenges. We see acts of profound care for each other that make us feel proud and honoured to be human beings. Then there are the other times – times when we attack each other and the planet that we live on; times when we override our normal feelings of responsibility for our actions; times when we could not be further from our position of sentient, intelligent, loving, caring, beings of compassion. Through our contact with these ETs, they are holding up a mirror to us all. It is time for all of us to take a long hard look in that mirror and ask the question: are we happy and content with what we see?

'The human race is on the verge of taking the next tentative step up the evolutionary ladder. To achieve this our DNA is changing, our vibrational rate is altering. This transition has been occurring for some time now, and the task is reaching completion. To ensure a healthy, untroubled labour during this birth, we must make a profound push towards the finishing line. Additional changes to our DNA and our vibrational rate will come from our own words and actions. What we say and do, every single moment of the day, will change who we are and will ultimately ensure our safe passage as life on this planet changes. It's not too late. There are agencies abroad that would dumb us down and would stymie our evolution, but I can assure you that they will not be allowed to stand in our way. It's entirely up to you now. You can decide to walk away from this day with a new outlook on your life. You can take responsibility for your words and actions. You can become a wonderful and empowered human being. You can realise that we are all the same – the human beings on this planet and life throughout the cosmos, that life is never ending and that we are beautiful, spiritual and unique beings of light.'

"Promise me you'll always remember:
You're braver than you believe,
and stronger than you seem,
and smarter than you think."

Christopher Robin to Pooh

The Awakening

I'm not a perfect human being; in fact I'm far from perfect, but who amongst us would want to be? After all, isn't this life of ours all about gaining wisdom through the process of living it with care and responsibility? So, if we were perfect, by definition that would suggest that the learning process was complete. We would have nothing else to aim for, and how dull would that be? I have discovered that life is not all about achieving perfection but rather improving the quality of how one lives one's life by responding to what is presented with awareness of how our words and actions affect everybody and everything around us. This discovery has now become the plain-and-simple recipe that allows me to try to live my life without the need to create destructive waves that rock the boats of others. It was always my hope that by reading about my experiences and how I responded to them, I might just be able to pass on a little support and comfort to those who have had the same experiences and to plant a few signposts for those who are on the path to their own personal reawakening.

In many ways, my life is unrecognisable to how it was nearly ten years ago; I have a wealth of friends and acquaintances connected to my 'contact awakening' and my world is blessed because of it. They have all played a part in helping me to not only come to terms with who I really am, but to move forward with it in a healthy and productive manner. I am a better person because of my ET contact and also because of the people I have met and the friends that remain loyally by my side. Sadly, there have been a few family members who have fallen by the wayside because they were unable to accept my reality of contact and what it has meant to Annie and me. I can only assume that their denial was down to their own personal fear of the subject matter or the restricted belief system that they have picked up along the way, for I have never been able to ascertain their true reasoning because of their aggressive and dismissive attitude. But that's okay, for the time will come – very soon in fact – where I will be able to sit down with those people and they will be able to listen with an open heart.

I suppose that what my contact has taught me – in greater detail – is another book on its own, but suffice to say that whatever I believe the contact to be about, with regards to the rest of the human race, it has fulfilled its goal to a lesser or greater degree with me.

I am waking up.

On every level imaginable, I am truly waking up.

Don't get me wrong, I don't believe that we will really know what the ET contact has been about – for them that is – but if we can take something from it that will enable us to reawaken our quiescent wisdom, then hallelujah brothers and sisters! For that is what I have been attempting to cram into my dense Neanderthal head for many life times: life never changes, but how one responds to what is presented can. And each life experience is just full of wisdom waiting to be harvested each time we go through it. There is truth, after all, in the concept of mind over matter, for with thoughts come words, which is normally followed by actions. And the whole process is manifested like a ripple on a pond as it touches the lives of all those around you. So the accountability is great my friends, for what we say and do does affect others and it is our responsibility to be aware of that and live accordingly. Also, the effect may not always be visible, for all action carries an energy that is never ending and may continue to have an impact forever more – even the simplest gesture of kindness, recognition, sharing or politeness. But there are always two sides to the coin, so also consider the words and actions of individuals from centuries past that still cause conflict in today's modern cultures through the differences expressed through belief systems and various religious factions.

The interesting thing about truth is that it has always been there – untouched by the world around it, just waiting for each and every one of us to discover and embrace it. One aspect of the truth is that human beings are extremely special – in many different and varied ways – but I have only just discovered it. That has been an incredibly important and vital part of my awakening. I have discovered that even through the bad times: the wars, the crimes, the earthquakes and the tsunamis; through the heartache, the poverty and the starvation; through the tears, the misery and the loneliness – I would not want to change who I am: a fragile, caring, sensitive, loving, flawed, intelligent, human being with the need and desire to live my life with authenticity and the awareness of its innate potential; and a Being that although small in its place in the cosmos, still remains aware of its larger place and impact in the grander scheme of all things.

The ETs (or whatever we decide to call them) remain in my life. Although the alien construction workers who laid the foundation of that contact might have receded and been replaced by another type of personage, more appropriate to this stage of my individual contact, we are all still together in this ongoing story. It might not play out like it once did, with night–time visits of small, spindly, alien-looking beings with large black eyes, restricted height and an apparently humorless demeanor – although I still sense their presence behind the scenery. What it has been replaced with now is something much more energetically delicate and refined – although I'm not suggesting that what came before was gross in anyway, just different.

Regrettably, my body has been irreversibly damaged, it would appear, through the actions of being taken. Why that is I do not know. Why they seem unable or reluctant to repair the damage, I have no idea. Maybe it isn't possible. They might believe it isn't their responsibility. Annie speculates that during the abduction process, when my body has been taken through solid objects and lifted up into the craft, whatever science/magic they use to carry out these procedures has altered my current physical and electrical construction. This has left me with legs that give me constant pain, that 'vibrate' at night uncontrollably (even when sleeping) making the muscle structure hurt as if I run a marathon every night. I also experience peculiar symptoms related, I believe, to the implants that have been inserted into various parts of my body and although I would not want them to be removed, I sense my human body is still attempting to reject them. My energy levels are lower than they should be, even at the ripe old age of 53, and I continue to have trouble sleeping, even though I no longer lie awake wondering when and if they are coming to take me out to play.

La–de–da! Hey–ho! Whatever! Que sera!

I know Annie has her theories and they might be true, but I have my own opinion.

I believe that things have been done to me that have changed the density and vibration of my body. I believe that I now find it difficult to function in the dense three-dimensional environment that is life on this planet. For the majority of the human race – although the figure is dropping constantly as the contact and the upgrades to the Contactees / Experiencers / Abductees continue – their density and construction works perfectly well for the world they live in, but very soon that will all be very different. That is why the alterations are being carried out for some of us, for when the shift occurs, when the human race takes the next tentative step up the evolutionary ladder and our home environment changes too, then we will have to be different, we will have to be less dense in construction and be able to move up and function in a higher vibrational realm. So, I believe my 'problems' originate from that and the alterations that have been made. When I think about it and ask the question in my head, I always get back the one word answer of 'electrical'. To this end also I have periodic 'downloads' of symbols that appear to go straight in to my brain or wherever this information might be stored. Once more I have been 'told' this is for the future 'upgrading' of my/our DNA codes.

It's as if the already wonderful vehicle that I use for transportation has been improved with a host of new applications, but still I'm putting in the wrong fuel and expecting it to operate properly in an environment now alien to its construction. Sometimes I feel like a fish out of water!

So that's why I believe my body moans and groans. I'm not suggesting that I'm ready for the knacker's yard quite yet, but I do need to manage my pain with extremely strong prescription painkillers, which is a shame after all and does get on top of me some days. Having said all that, you need to know that I have never been happier than I am right now. My life is full of joy. I am more comfortable being in my own skin, just being me, than ever before. My relationship with Annie has improved because of my awakening, because I have finally and totally ascribed to the set of guidelines that I am trying to share with everybody else – lead by example so they say. I don't suppose I am ever going to be the perfect human being – whatever that is – but at least now I am a much more refined role model. I believe that I am now closer to being ready for the next stage, the next step that will fulfill our human destiny.

I would also like to say that there have been a number of other 'changes' in me that are worth mentioning of a much more positive nature. My psychic abilities have been greatly enhanced because of my ET contact; I say enhanced because I know that I was born with these abilities (as I have explained in a previous part). Although 'society' had a bloody good try squashing them down as I grew up, there was obviously a sympathetic and supportive agency at work to defend that manipulative containment. Therefore, when it became appropriate for me to have full working access to the 'toy box', it came as second nature to me to once more work outside of the normal five senses.

As a Reiki practioner, I quickly discovered that I was not working within the accepted parameters of the individual therapies that I was practicing. In fact, whatever I was doing when it came to the healing of my clients, it was not coming out of the teaching sessions I had attended or the training books I had read. I was drawing upon an energy far greater than I was aware of and in doing so I was fulfilling part of my life objective to help other people. It is a privilege to work with this life energy and an honour that my clients put their trust in me. I have literally seen – what I would consider – miracles happen when I have worked with some people. Please don't get me wrong, I am merely a conductor for this healing energy and I do not believe that I have any direct force or effect upon it. The energy is naturally there, it is processed through me and it is the decision of the person I am working with if they accept it. If it hadn't been for my ET contact,

I may not have been doing it.

As for my fellow human beings, I see through their eyes more often now and watch myself in closer detail, which obviously assists in my continued growth of awareness. I can recall the moment this amazing transition began: it was the first time I was able to see life through my son's eyes. My dear son, my lovely Reecey–Boy stood there in the turmoil of growing pains, amidst the crippling anguish of transition between childhood and maturity and cried out for help. It was at that moment I was able to be more than his Dad, it was then I was able to experience his dilemma through his eyes, because I finally remembered I had been there once myself. No two lives or their experiences are the same, but at least I was able to tune back into my troubled youth and re-experience what I had gone through and in doing so I was able to refine the love and care I was able to offer him. I truly believe that in doing so I was able to pass on some of that wisdom to him because I see now that Reece has finally begun to find himself; at last he appears to be – just like his old Dad – happier and more comfortable being in his own skin. His wisdom though is his own and that's how I would want it to be. And he has, without probably even knowing, blessed my life in a million different ways and given me *a magical pocket full of keys*.

This life is swings and roundabouts, but now I recognise that wisdom comes through all experiences, so it is important – nay vital – to partake of all of the carnival attractions: the swings, the roundabouts and everything else that is on offer. I am glad to be alive and I really do hope to be alive for a lot longer. I love me, I love my life and I love my world – what I have created and what it is.

Do you think it might just be possible though to blame my hair–loss on my abduction experiences? No? Oh well, it was worth trying.

I am in constant contact with the beings that I have referred to as The Light Beings. Unlike the worker bees that are The Greys, The Light Beings work in a different way. I don't believe that they are three-dimensional as we would understand it, so when they actually focus in and communicate with me, it is with my light body, my spirit, my soul – although I am still able to see them with my physical eyes on the very few occasions when they have chosen to manifest in that way. For the rest of the time I can sense their presence. It's as if the phone has been left off the hook and my thoughts, feelings and emotions trickle out into wherever they are, just like my spiritual ears can hear them at the end of the line. When my consciousness silently raises a question, they answer it without me actually asking or really being aware the response is coming back from them. Maybe a better way of describing it would be to say that the ETs are *in here with me*. The Greys used to respond to me – when they weren't physically in my presence – via my implants (and may still do for all I know), but it's much more sophisticated with The Light Beings; it feels as if a part of me now belongs to them and somehow resides in their state of existence at the same time.

Am I still having contact experiences? Absolutely. It has always been a major part of my life, and although it has changed, it remains in place. I was never really sure why I was supposed to write about my contact; now I understand completely. My response to my experiences and my input to the larger picture of world-wide contact via this book is merely a tiny piece of the puzzle of what 'contact' has always meant to all of us. Although only a *piece* of the puzzle, without it the picture would be incomplete. Also, I acknowledge that my existence on this planet is a contribution to the bigger picture too and what I say and do on a daily basis will continue to define my purpose and reflect back what our continued path should be and what our joint goal truly is. *It has always been about the awakening of the human race.* It is about us taking responsibility for our words and actions, treating each other – every sentient being on the planet – with love and care and tending our home, the beautiful, unique and wonderful planet Earth, with the respect it deserves.

We have slept for so long, but now it is time to wipe the sleep from our eyes, wash our faces, comb our hair and put on our best party clothes, for it is time to take up our *invitation to the dance*.

CPSIA information can be obtained at www.ICGtesting.com
229408LV00002B/2/P